Evaporation and Evapotranspiration

Wossenu Abtew • Assefa Melesse

Evaporation and Evapotranspiration

Measurements and Estimations

Springer

Dr. Wossenu Abtew
South Florida Water Management District
West Palm Beach, FL
USA

Prof. Dr. Assefa Melesse
Department of Earth and Environment
Florida International University
Miami, FL
USA

ISBN 978-94-007-4736-4 ISBN 978-94-007-4737-1 (eBook)
DOI 10.1007/978-94-007-4737-1
Springer Dordrecht Heidelberg New York London

Library of Congress Control Number: 2012946197

Printed on acid-free paper

Springer is part of Springer Science+Business Media (www.springer.com)

Preface

Water loss through evaporation from open water and evapotranspiration (ET) from vegetation is one of the major components of the hydrologic cycle affecting water resources availability. Measurement and estimation of these terms have initiated the development of the theory of the process, measurement techniques, and estimation equations. Perceptions have contributed to biases of estimation. A drying pond taken over by vegetation gives the perception that the vegetation's increased ET resulted in the drying of the pond. Succession of vegetation in a wetland may hide the impact of changing hydrology by suggesting water losses are due to invading vegetation.

The evapotranspiration process is controlled by the availability of moisture to evaporate. Energy is required to detach water molecules. A mechanism is required to move the vapor into the air column. The air has to have the capacity to hold the vapor. When the air has no more capacity to hold moisture, the reverse process, dew formation, occurs. In this book, dew evaporation is presented in a chapter. A chapter on vapor pressure and vapor pressure deficit estimation methods is presented with known quality data from a monitoring network. ET processes and mechanisms are presented in a simplified way without compromising complexity. In each case, examples of applications from the authors' experience are presented for comparing estimation methods. Meteorological monitoring and data quality, input into ET estimation methods, is vastly discussed in a chapter with illustrations from a large monitoring network. The design and application of a lysimeter system for open water evaporation and wetland vegetation ET has provided measured data to gauge the performance of various estimation equations. The advantage and limitation of simple ET estimation methods, when input data is limited, are addressed. Remote sensing application to ET estimation is sufficiently addressed in three chapters with application case studies. An introduction into the expected impact of climate change on ET rates is included as a chapter with climate model application results. This book is a useful resource for hydrologists, scientists, meteorologists, engineers, water resource managers, agricultural and environmental professionals, students, and teachers.

Wossenu Abtew

Contents

Symbols and Abbreviations

A	Area
a, b	Coefficients
AET	Actual ET
a_w, b_w	Coefficients
c	Adjustment factor
c_1, c_2, c_3, c_4	Coefficients
C_{et}	Reference crop coefficient
C_n, C_d	Coefficients
c_p, C_p	Specific heat of air, heat capacity of air
c_s	Soil or water heat capacity
d	Displacement height
d_{TM}	Constant (0.1238 $mWcm2sr^{-1}\mu m^{-1}$)
de	Change in vapor pressure
$d_{e\text{-}s}$	Relative distance between Earth and Sun in astronomical units
DN	Digital number
d_r	Inverse squared relative distance between Earth and Sun
d_s	Effective depth
dT	Change in temperature between two measurement heights
ΔT	Change in temperature
dt	Change in time
du	Change in wind speed
d_w	Water depth
dz	Change in wind speed measurement height
e	Errors
E, LE	Vapor flux, latent heat flux, evaporation
e_d	Actual vapor pressure
e_{dd}	Vapor pressure in the air above evaporating surface
elev	Elevation above sea level
E_L	Lake evaporation
E_o	Open water evaporation
e_o	Saturation vapor pressure at lake surface

E_p	Potential evaporation
E_{pan}	Pan evaporation
e_s, e_a	Saturation vapor pressure
e_{ss}	Vapor pressure at evaporating surface
ESUN	The mean solar exoatmospheric irradiance
ET	Evapotranspiration
ET_{24}	Daily ET from remotely sensed instantaneous ET
ET_{aero}	Aerodynamic component ET
ET_c	Actual crop evapotranspiration
ET_{frac}	ET fraction for each pixel (average of hot and cold pixels)
ET_i	Remotely sensed instantaneous ET
ET_o	Evapotranspiration from grass reference crop (8 to 15 cm and well watered)
E_p	Potential evaporation
ET_p	Potential evapotranspiration
ET_r	Reference crop evapotranspiration; grass reference ET
ET_{rad}	Radiation component ET
ET_{ref}	Reference ET
ET_r F	Alfalfa reference evapotranspiration fraction
ET_{sz}	Standardized reference crop evapotranspiration for short or tall crop
f	Fractional vegetation cover
$f(u)$	Function of the horizontal wind
F_c	Fraction of cover
f_c	Fractional canopy cover
G	Heat storage
GAIN	Solar spectral radiance for each band
g_b	Boundary layer conductance
g_c	Canopy conductance
g_m	Measured conductance of leaf
g_s	Stomatal conductance in mmol m^{-2} s^{-1}
G_{sc}	Solar constant
g_{sv}	Stomatal conductance in mm s^{-1}
H	Sensible heat
h	Reference vegetation height
h_c	Average height of cover or crop height
H_s	Sensible heat for soil surface
H_v	Sensible heat for vegetation surface
I	Inflow
J	Julian day
k	Von Karman constant
K_1, K_2, K_3	Coefficients
K_{1ls}, K_{2ls}	Calibration constants for Landsat 5 and 7
K_c	Crop coefficient
k_h	Coefficient for sensible heat transfer
k_m	Mass transfer limiting term

K_p	pan coefficient
K_t	Transfer coefficient
k_w	Coefficient for latent heat transfer
L	Obukhov length
LAI	Leaf area index
LE_s	Latent heat for soil surface
LE_v	Latent heat for vegetation surface
L_j	Leaf area index for canopy strata j
L_{max}	Maximum spectral radiance
L_{min}	Minimum spectral radiance
m	Constant (0.0056322 $mWcm^2sr^{-1}\mu m^{-1}$)
MSE	Mean square error
n/N	Mean actual to possible sunshine ratio
NDVI	Normalized Difference Vegetation Index
NDVs	Scaled NDVI
NIR	Near infrared band
N_o	Mass transfer coefficient
NTC	Negative temperature coefficient
O	Outflow
P	Atmospheric pressure
p	Mean daily percentage total annual daytime hours
PRT	Platinum resistance thermometer
PTC	Positive temperature coefficient
q	Specific humidity
q'	Specific humidity fluctuation
Q_a	Advective energy gain or loss
Q_h	Sensible heat gain or loss
Q_{in}	Energy input into the system
Q_{out}	Energy leaving the system
Q_{Rn}	Energy from net solar radiation
r	Correlation coefficient
R	Linear function of the digital number (DN)
R_A	Extraterrestrial solar radiation
r_a	Aerodynamic resistance
R_b, R_L	Net back or outgoing thermal radiation
r_c	Canopy resistance
RED	Red band
R_f	Rainfall
RH	Relative humidity
RH_{avg24}	Average humidity from 24-h continuous observations
RH_{max}	Daily maximum relative humidity
RH_{min}	Daily minimum relative humidity
r_l	Stomatal resistance of a single leaf
R_n, R_{Sn}	Net solar radiation

$R_{n,s}$	Net solar radiation on soil surface
$R_{n,v}$	Net solar radiation on vegetation surface
R_s	Incoming solar radiation
r_s	Stomatal resistance
R_s	Resistance to heat flow in the boundary layer immediately above the soil surface
R_{so}	Clear sky solar radiation
R_x	Ground reflectance for band x
S	Slope
S_{cj}	Stomatal conductance of leaf strata j
Sp	Seepage
Std	Standard deviation
$S_{y/x}$	Standard error
T	A given temperature
T_a	Air temperature over a lake, near surface air temperature
T_{avg}	Average air temperature
T_{avg24}	Average temperature from 24-h continuous observations
T_d	Dew point temperature
T_{max}	Daily maximum air temperature
T_{min}	Daily minimum air temperature
T_n	Average temperature on day n
T_{n-1}	Average temperature on previous day
T_s	Lake surface water temperature
T_{sur}	Radiometric surface temperature
T_v	Vegetation surface temperature
$u*$	Friction velocity or shear velocity
u_{day}	Daytime wind speed
u_z	Wind speed at height z
vpd, δe	Vapor pressure deficit
w	Vertical wind speed
w'	Vertical wind speed fluctuation
WI	Wetness index
z_h	Roughness length for heat transfer
z_o/z_{om}	Aerodynamic roughness/ roughness height or length for momentum transfer
z_{oh}	Roughness length for vapor and heat transfer
α	Albedo
$\alpha_{path\text{-}radiance}$	Path radiance albedo
α_{toa}	Albedo of the top of atmosphere
β	Bowen ratio
γ	Psychrometric constant
δ	Change in depth
$\varDelta$	Slope of vapor pressure curve
Δe	Change in vapor pressure
ΔQ_s	Change in energy storage

ΔS	Change in storage
ΔSM	Change in soil moisture
ΔT	Change in temperature with time
ε	Ratio of molecular weight of water to dry air
ε_a	Atmospheric emissivity
ε_s	Surface emissivity
ζ_{short}	Absorptivity
Θ^*	Temperature scale
θ, sδ	Solar declination angle in radians
θ_a	Potential air temperature at height z
θ_o	Potential temperature at the surface
$\theta_o - \theta_a$	Mean surface temperature
θ_v	Potential virtual temperature near the surface
λ	Latent heat of vaporization of water
Λ_r	Relative evaporative fraction
λ_s	Thermal conductivity of soil
ρ	Air density
Γ_c	Coefficient
Γ_v	Coefficient
σ	Stefan–Boltzmann constant
τ	Shear stress
τ_o	Surface shear stress
τ_{sw}	One-way atmospheric transitivity
φ	Latitude in radians
Ψ_h	Stability correction factor/function for sensible heat transfer
Ψ_m	Stability correction factor/function for momentum transfer

Chapter 1
Introduction

1.1 Overview of Evaporation and Evapotranspiration Studies

Evaporation from open water and wet surfaces and evapotranspiration from vegetation are one of the major parameters in the hydrologic cycle. Most precipitation is lost in the form of evaporation and evapotranspiration with the percentage varying from region to region globally. Spatial variation by latitude, longitude, altitude, environment, and specific site conditions is a source of variation in evaporation and potential evapotranspiration. Standardized measurement and estimation of this parameter are challenging. Even with estimation methods standardized, variation in estimates would occur due to lack of uniformity in input data collection and quality control. A positive characteristic of this parameter is that it has relatively smaller variation for a given time and location. Seasonal fluctuations are known, and ranges are limited when water is not a limiting factor. Estimation error is relatively lower if appropriate equation and good quality input data is used for a given site.

Apart from individual publications on the subject, the United Nations Food and Agriculture Organization (FAO) and the American Society of Civil Engineers (ASCE) have made major contributions toward developing common understanding of the science of evaporation and evapotranspiration and standardizing estimation methods. ASCE's consumptive use of water and irrigation requirements (Jensen 1973) provided the most detailed information on evapotranspiration for that period with various evapotranspiration and potential evapotranspiration estimation methods documented and evaluated. FAO Irrigation and Drainage Paper No. 24, Crop Water Requirements (Doorenbos and Pruitt 1977) presented the Blaney–Criddle, Penman, and pan evaporation methods for estimating reference crop evapotranspiration. The presentation is organized for wide-scale application with tables and charts. Crop coefficients are provided for various crops. Irrigation scheduling guidance is also provided with application rate estimations. ASCE manual and reports on engineering practice No. 70, Evapotranspiration and Irrigation Water Requirements

W. Abtew and A. Melesse, *Evaporation and Evapotranspiration: Measurements and Estimations*, DOI 10.1007/978-94-007-4737-1_1,

(Jensen et al. 1990) build on past publications and provide details on methods and parameter estimations. Methods are evaluated using lysimeter-measured evapotranspiration data from various locations. Distinction is presented between potential and reference evapotranspiration. Crop coefficients are provided for various crops including varying by stage of crop growth. The ASCE Standardized Reference Evapotranspiration Equation (Allen et al. 2005) was published for standardizing reference evapotranspiration estimation methods by providing a single equation with common procedures to derive or estimate certain inputs. The American Society of Agricultural and Biological Engineers dedicated periodic conferences on evaporation and irrigation scheduling producing conference proceedings that contributed to the advancement of the science and application to agriculture.

This book builds on existing works on the subject but introduces a fresh and new approach. Topics as lake evaporation and wetland evapotranspiration have not been given this scale of analysis in the past. Each is presented in a chapter. Lysimeter measurements are used to demonstrate application of various methods. New simplified equations are presented. Estimation of evapotranspiration depends on the quality of input data. This book sufficiently covers meteorological monitoring and data quality based on experience of meteorological data collection. The quality of evapotranspiration estimates is dependent both on the selected model and the quality of the input data. While most publications intensively evaluate ET estimation equations, input data quality is not sufficiently evaluated. In most cases, input data sources are external, and data quality is not available with the data. In this book, a chapter is devoted to meteorological parameter monitoring, sensors, challenges in acquiring good quality data, and data quality evaluation. Illustrations of poor and good data quality are provided.

Evaporation and evapotranspiration measurements are presented from the simple pan to remote sensing methods. Remote sensing application to evapotranspiration quantification is covered in three chapters covering presentation of the various surface energy balance models utilizing remotely sensed data, application of remotely sensed based ET for crop yield estimation, and also evaluation of wetland restorations. Case studies demonstrating the application of remote sensing are presented.

Dew formation and the energy required to evaporate dew are presented in a chapter with results from experimental work. Energy balance and mass transfer during early morning dew evaporation are discussed in full detail. A chapter is devoted to evaluation of many types of vapor pressure calculation methods using quality-controlled meteorological data collection. A review of global warming and climate change projected impact on rates of evapotranspiration is explored in a chapter with literature review and model applications.

There are global, regional, local, and site-specific evapotranspiration estimation products provided by commercial, governmental, and academic institutions. In several cases, graphic and digital products are provided with not much explanation in what equations and data were used to generate the product. Nevertheless, the products satisfy various needs. A global average actual evaporation monthly product developed from meteorological data input of 1985 to 1999 based on the JULES model is available on the web (http://www.jchmr.org/jules. Accessed on 13 October 2011). JULES output is based on simulation of evaporation from soil and canopy as well as the surface of lakes, wet vegetation canopies, and snow. High evaporation in May in the northern hemisphere and in February in the southern hemisphere is shown. It illustrates globally that evaporation is limited by moisture availability and by the variables that affect evaporation.

Annual average potential evapotranspiration estimates for the continental United States are posted on the web (http://serc.carleton.edu/images/introgeo/socratic/examples/USevapotran.jpg. Accessed 13 October 2011). Global monthly evaporation total and anomalies are provided by the National Oceanic and Atmospheric Administration (http://www.cpc.ncep.noaa.gov/cgi-bin/gl_Evaporation-Monthly.sh. Accessed 05 December 2011). A reference evapotranspiration map for Africa is provided by the Food and Agriculture Organization (FAO) of the United Nations. The reference evapotranspiration data is derived using the FAO Penman–Monteith method as described in FAO Drainage Paper 56 (Allen et al. 1998) at a web site (http://www.fao.org/nr/water/aquastat/watresafrica/index3.stm. Accessed 13 October 2011).

Annual areal potential evapotranspiration estimates for Australia are provided by the Australian Bureau of Meteorology. Areal evapotranspiration is computed based on Morton's (1983) complementary relationship areal evapotranspiration model. Morton's model for areal evapotranspiration is a modified Priestley–Taylor equation with modification for advection (Wang et al. 2001). Caution is added in the documentation that the ET map is subject to error from input data measurement error, sampling error, interpolation and mapping error, and model error. The map can be accessed on the web (http://www.bom.gov.au/jsp/ncc/climate_averages/evapotranspiration/index.jsp?maptype=3&period=an. Accessed 13 October 2011).

Evaporation and evapotranspiration estimation is presented in this book in a chapter in detail with application of simple to complex models. Model comparison is presented with input meteorological data of known quality. Selection of the best estimation model for a location with limited input data sets is made simpler. Reference and crop evapotranspiration is presented in a chapter making the link between reference evapotranspiration and actual crop evaporation. All 13 chapters contain valuable material for reference and applications.

References

Allen RG, Pereira LS, Raes D, Smith M (1998) Crop evapotranspiration – guidelines for computing crop water requirements. FAO irrigation and drainage paper 56. FAO, Rome

Allen RG, Walter IA, Elliott R, Howell T, Itenfsu D, Jensen M (2005) The ASCE standardized reference evapotranspiration equation. ASCE, Reston

Doorenbos J, Pruitt WO (1977) Guidelines for predicting crop water requirements. FAO irrigation and drainage paper 24. FAO, Rome

Jensen ME (1973) Consumptive use of water and irrigation water requirements. ASCE, New York

Jensen ME, Burman RE, Allen RG (eds) (1990) Evapotranspiration and irrigation water requirements. ASCE manuals and reports on engineering practice No. 70. ASCE, New York

Morton FI (1983) Operational estimates of areal evapotranspiration and their significance to the science and practice of hydrology. J Hydrol 66:1–76

Wang QJ, McConachy FLN, Chiew FHS, James R, de Hoedt GC, Wright WJ (2001) Climatic Atlas of Australia maps of evapotranspiration. Australian Government, Bureau of Meteorology, Melbourne

Chapter 2
Meteorological Parameter Monitoring and Data Quality

Abstract Evaporation and evapotranspiration (ET) estimation models require meteorological observations as input. The quality of the input data determines the quality of evaporation and evapotranspiration estimation. It is important to discuss the subject of meteorological monitoring, types of sensors, and challenges of operation and maintenance. In this chapter, the most common meteorological variables and examples of respective sensors are presented. Examples of meteorological data quality variations that reflect characteristics of most monitoring systems are presented. The significance of input meteorological data quality in determining the quality of ET estimates is addressed.

Keywords Meteorological data • Data quality • Solar radiation • Wind speed • Temperature • Humidity • Air pressure

2.1 Introduction

Evaporation and evapotranspiration estimation models require input data that are field observations, derived or assumed parameters. Field measurement of meteorological variables is a critical part of the evaporation estimation process. Measurements and recording errors in field variables result in evaporation and ET estimation errors. In this chapter, current instrumentation for meteorological variables observations is presented. Potential shortcomings in data quality are illustrated with actual field observations from a large monitoring network.

2.2 Meteorological Parameters Monitoring Network

In general, the objective of hydrometeorological monitoring is point measurement of temporal variation of each variable. Monitoring network design principles has not been practiced very well. As a result, in many places, haphazard networks

W. Abtew and A. Melesse, *Evaporation and Evapotranspiration: Measurements and Estimations*, DOI 10.1007/978-94-007-4737-1_2,

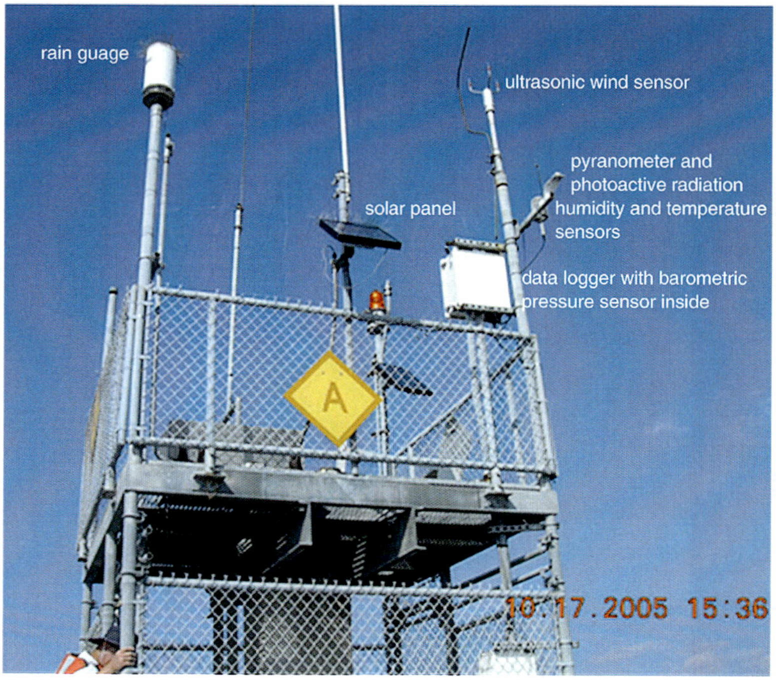

Fig. 2.1 A weather station in Lake Okeechobee, south Florida (Photograph provided by South Florida Water Management District)

exist as a product of placing monitoring sites without network design with a site or group of sites selected for specific objectives. Ideally, network design should come before monitoring site selection and instrument installation. The result has been too many monitoring sites in some areas and sparse in other areas. Point measurement is spatially assumed to represent areas that are closest to the point of measurement. Depending on spatial variation of a parameter, a monitoring site may represent a large area. Monitoring point location selection has not been given careful consideration as demonstrated by many historical and current sites where locations are subject to interferences. Historically, purposes of monitoring site selection, sensor installation, and operation could differ from site to site, and more than one institution could be involved in the expansion of the local network. Differences in data quality reflect differences in site management.

Most meteorological variables have known ranges. Deviations of measurements from the "true" value are a function of instrument type, operation and maintenance, personnel skill, maintenance plan, data recording, transmission, and storage capability. The most common meteorological parameters are air temperature, humidity, barometric pressure, solar radiation, net solar radiation, and wind speed. Wind direction, photoactive radiation, leaf wetness, and water temperature are additional parameters measured that may not be common in all weather stationss. Figure 2.1 depicts a weather station tower inside Lake Okeechobee, south Florida, with various sensors shown.

2.2.1 *Sources of Error in Meteorological Parameters*

Errors can be classified into three categories: systematic, random, and process errors. Systematic errors are caused by sensor manufacturing defects or calibration error putting constant upward or downward drift in observations. Once these types of errors are indentified, correction can be made to the sensor. Some data can be salvaged through application of correction factors. Random errors are errors whose sources may not be known and occur in both increasing and decreasing direction without discernable pattern. Sources of errors are instrument malfunction, instrument limitations, instrument calibration and programming, environmental factors, data recording and transfer, data processing, and storage. Process errors are those incurred through data recording, processing, transfer, and storage.

Errors can be reduced by application of quality control measures through the development of standard operating procedures for field data collection, data processing, and storage. The quality of staff training reflects on data quality. Proper instrument installation, calibration, testing, and regular maintenance are required to collect data with acceptable quality. Data error detection guidelines/software, error reporting, correction, and remediation processes minimize the rate of data error. Peer review processes and frequent publication of data contribute to improvement in data quality and make necessary corrections before long periods of erroneous data collection.

2.3 Air Temperature

Air temperature is one of the easiest parameters to measure. Commonly, decent quality air temperature data is available in many regions. The availability of multiple sources of observations for one locality provides the ability to evaluate the quality of a data set from a single site. There are numerous models and types of air temperature sensors. Each type has different ranges of observation and accuracy and measurement error. Air temperature gauges include platinum resistance thermometer (PRT) and thermistors and internal conductance temperature probes (Crowell and Mtundu 2000). The platinum resistance thermometer is placed in a ventilated air chamber to shield the sensor from radiated heat energy. There are two types of thermistors. A positive temperature coefficient (PTC) thermistor increases resistance with increase in temperature. A negative temperature coefficient (NTC) thermistor decreases resistance with increase in temperature. The internal conductance temperature probe is a semiconductor device with internal conductance changes that are proportional to temperature. A signal conditioning unit processes the probe signal to provide a 0–5 V output that is directly proportional to temperature. Measurement accuracy for the PRT is ± 0.05°C. A contemporary air temperature gauge is the HMP45C temperature and relative humidity probe. The range is -33 to 48°C with accuracy of ± 0.4°C over the full range (Abtew et al. 2007). Figure 2.2 depicts an air temperature and humidity probe and the ventilator housing.

Fig. 2.2 HMP45C air temperature and relative humidity probe with data logger (Photograph provided by South Florida Water Management District)

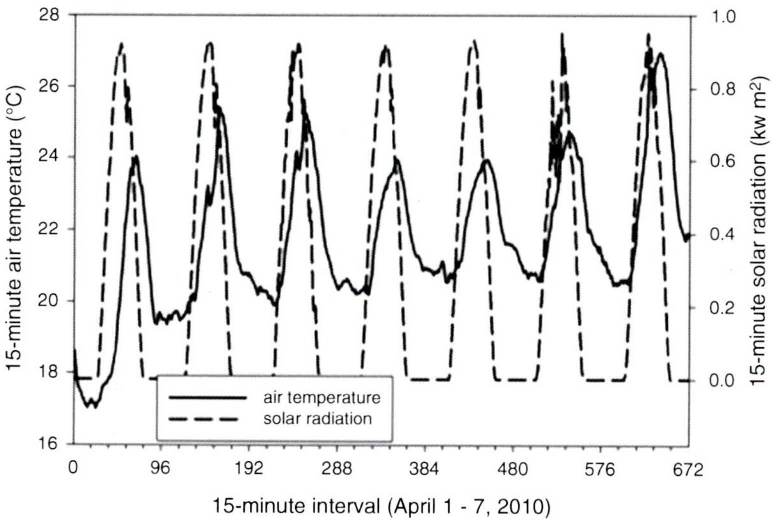

Fig. 2.3 Diurnal fluctuations of air temperature and solar radiation in south Florida (April 1–7, 2010)

Seasonal and diurnal fluctuations in air temperature for a location are well known making data quality evaluation relatively easy. Diurnal variation of air temperature is caused by the characteristics of solar and terrestrial radiation. Generally, in the daytime, incoming solar radiation is higher than outgoing terrestrial radiation and vice versa at night. This results in warmer temperature at daytime and cooling at nighttime due to the rotation of the Earth. Figure 2.3 depicts 15-min air temperature and solar radiation for 1-week period (April 1–7, 2010) at the L006 weather tower inside Lake Okeechobee in south Florida. The diurnal fluctuation of air temperature

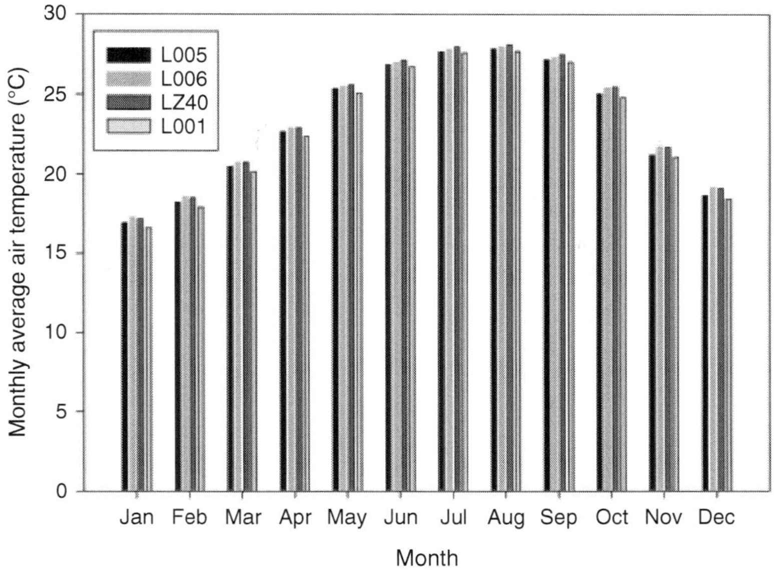

Fig. 2.4 Comparison of monthly average air temperature from four sites in Lake Okeechobee, south Florida (1994–2010)

Table 2.1 Comparison of daily air temperature (°C) from four sites in Lake Okeechobee (1994–2010)

Site	L005	L006	LZ40	L001	Overall
Mean	23.21	23.49	23.53	22.99	23.28
STD	4.62	4.40	4.51	4.65	4.55
Overall deviation	−0.07	0.21	0.25	−0.29	0

correspondence with solar radiation is clearly shown with the peak of solar radiation lagging behind the peak of air temperature. Since a week with a complete data set was selected for this analysis, a data quality problem was not observed.

Seasonal variation of air temperature follows seasonal variation of solar radiation in an annual cycle. The Earth revolves around the Sun once each year. Since the axis of the Earth is tilted by 23.5°, the angle of incident of solar radiation changes seasonally with the Northern Hemisphere being warm in June, July, and August and the Southern Hemisphere in December, January, and February. Figure 2.4 depicts seasonal variation of air temperature in south Florida from four weather stations inside Lake Okeechobee for 16 years (1994–2010). Lake Okeechobee located at 27° latitude and 81° longitude has a surface area of 1,732 km^2 and mean depth of 2.7 m (Jin et al. 1998). Table 2.1 depicts a 5-year statistical summary of air temperature observations at the four sites. As shown in Table 2.1 and Fig. 2.4, the spatial variation and possible differences in temperature measurement between the four gauges are small. Data quality for air temperature at the four sites for the period of analysis includes days with missing data, days with partial observations (for part of the day), and days with estimated data. Missing or partial observation days for a site were filled with the average of the remaining sites where data is available. The small variation between the four sites could be due to small spatial variation.

Latitudinal variation of air temperature is related to latitudinal variation of solar radiation. Because of the near spherical shape of the Earth, the Sun is nearly overhead around the equator, and the angle of incidence decreases from 90° with increasing latitude. Seasonal variation of solar radiation and temperature in equatorial region is smaller than at higher latitudes. Similarly, day length is constant at the equator but varies seasonally with increasing latitudes.

Air temperature decreases with increasing altitude in the lower atmosphere (troposphere) for given latitude. This phenomenon is known as the environmental lapse rate. The average decrease is 6.5°C per 1,000-m altitude rise. Air temperature measurement is sensitive to the environment the sensor is exposed to. The general recommendation is to place the sensor at 1.2–2.1 m above ground level exposed to unobstructed air conditions at the site (WMO 1996). The site has to be exposed to sunshine and wind without obstructions such as buildings and trees.

Sources of error in air temperature measurement are due to improper installation and instrument malfunction. The site of the temperature sensor must be representative of the area of interest. The observed temperature should be representative temperature of the free air condition. Improperly installed gauges on unrepresentative surfaces, or sheltered by obstruction, may not provide unbiased observations. Concrete pavement sites would be subject to source of advective heat unless the objective of measurement is for such an area.

2.3.1 Dew Point Temperature

Dew point temperature is the temperature to which the air must be cooled to reach vapor saturation and result in condensation. Dew point temperature is established using a PRT embedded in a metal mirror. The mirror temperature is cooled to a temperature where condensation of vapor results on the mirror surface from the surrounding air. The temperature at this point is the dew point temperature. In Chap. 4, Energy Requirements of Dew Evaporation, observations of dew formation and evaporation are discussed in details.

A type of dew point temperature sensor is the General Eastern hygrometer with temperature sensor. Air flows convectively through a sensor chamber containing a dew point sensor and an air temperature sensor. The temperature of the metal mirror is controlled by a Peltier effect device. This is a device that heats when current is passed in a forward direction and cools when current flows in the opposite direction. The direction of current flow is determined by a light-emitting diode pointed at angle at the mirror surface (Crowell and Mtundu 2000). Wet-bulb and dry-bulb temperature dew point sensors, sling psychrometer, have 0.5°C accuracy, and

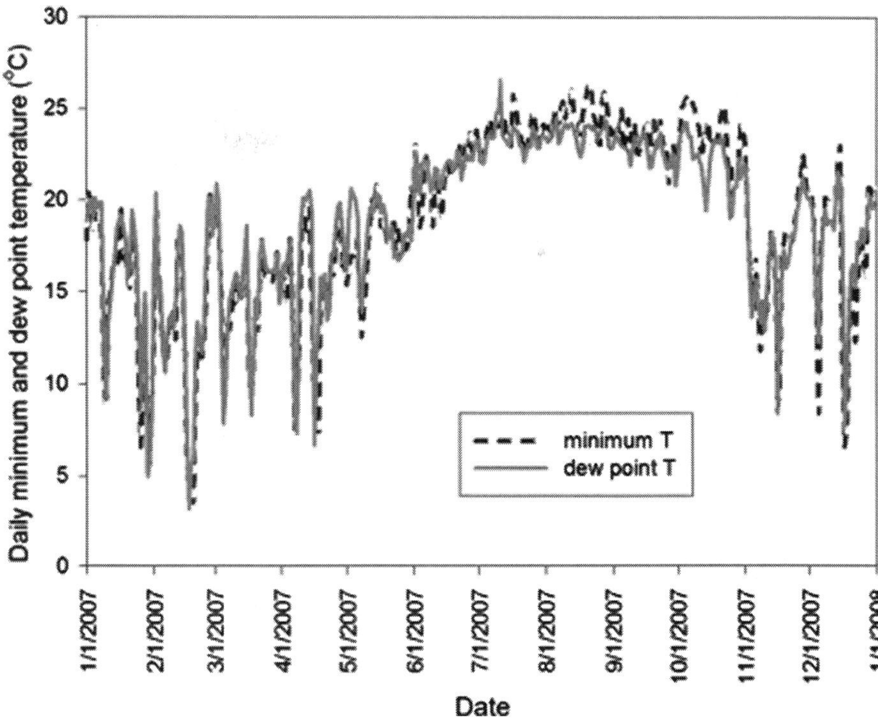

Fig. 2.5 Observed daily minimum air temperature and estimated dew point temperatures in south Florida

aspirated hygrometer has 0.1°C accuracy. In the absence of observed data, equations have been developed to estimate dew point temperature. Equation 2.1 (Bosen 1958) and Eq. 2.2 (Allen et al. 2005) give results with small differences:

$$T_{\mathrm{d}} = \left(112 + 0.9T_{\mathrm{avg}}\right)\left(\frac{\mathrm{RH}_{\mathrm{avg}}}{100}\right)^{0.125} - 112 + 0.1T_{\mathrm{avg}} \tag{2.1}$$

where T_{d} is dew point temperature (°C), T_{avg} is average air temperature (°C), and $\mathrm{RH}_{\mathrm{avg}}$ is average relative humidity in percent.

$$T_{\mathrm{d}} = \frac{116.91 + 237.3\ln(e_{\mathrm{d}})}{16.78 - \ln(e_{\mathrm{d}})} \tag{2.2}$$

where e_{d} is actual vapor pressure (kPa). Comparison of daily minimum air temperature and estimated dew point temperature for 2007 in south Florida is depicted in Fig. 2.5. As it is apparent, the daily minimum air temperature is a good dew point temperature estimate for south Florida. This may not be the case at higher altitude and latitude regions with lower humidity.

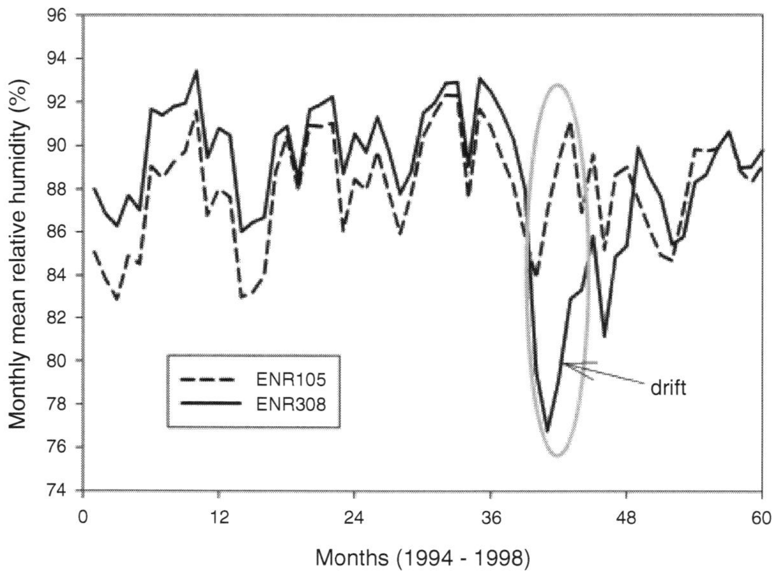

Fig. 2.6 Drift of a relative humidity sensor

2.4 Humidity

The air holds water in the form of water vapor from evaporation from wet surfaces and evapotranspiration from vegetation. The water vapor holding capacity of the air decreases with decreasing air temperature reaching saturation. When temperature lowers down to the dew point, the vapor in the air condenses into liquid water. Relative humidity is the measure of vapor amount in the air where 100% corresponds to saturation and lower percentages indicate drier conditions. Vapor pressure (kPa) is also a measure of vapor content of the air. Saturation vapor pressure is the measure of maximum vapor holding capacity of the air at the prevailing temperature. Actual vapor pressure is a measure of vapor content of the air at a given time.

Combined air temperature and relative humidity sensors include the Sierra–Misco model 2046 relative humidity and temperature sensor. This is a solid state device that measures temperature with an integrated circuit and humidity with a cellulose crystal. Accuracy of electric hygrometers is 2%. The HMP45C temperature and relative humidity probe contains a Vaisala HUMICAP 180 capacitive relative humidity sensor. At 20°C, the accuracy is $\pm$ 2% between 0 and 90% and $\pm$ 3% above 90% humidity. Humidity sensors are subject to failure due to instrument malfunction and other causes. Comparison of two sets of monthly mean relative humidity from two sites, 5.5 km apart, in a constructed marsh in south Florida shows substantial drift of one gauge for a few months (Fig. 2.6). Reported observations higher than 100% were replaced with 100%.

Fig. 2.7 A Campbell Scientific thermistor water temperature probe installed in Lake Okeechobee, south Florida (Photograph provided by South Florida Water Management District)

2.5 Water Temperature

Water temperature measurements at different depths provide a temperature profile along the depth of a lake. Campbell Scientific models 107 and 108 are examples of sensors used for water or soil temperature measurement. The probes consist of a thermistor encased in an epoxy-filled housing. The housing protects the thermistor and makes submerged installation possible. The probe measures typically at an accuracy level of $\pm 0.5°C$. A Campbell Scientific thermistor temperature probe installed in Lake Okeechobee is shown in Fig. 2.7 (Abtew et al. 2007).

Water temperature observations at three depths were made in Lake Okeechobee at weather station L006. The top probe was at 15 cm below water surface, the middle probe was at the middle of the lake depth, and the bottom was 30 cm above the floor. The average depth of the lake at the site is 2.78 m. Figure 2.8 depicts 3 years of water temperature observations at three depths.

These temperatures are unique to a subtropical shallow lake in a warm region. Deeper lakes at higher altitudes and latitudes will display different profiles based on the season including showing distinct stratification and over turning. Diurnal fluctuation of water temperature is less than that of air temperature. Data quality issues for water temperature from the monitoring system used in this analysis mostly deal with missing data and estimated data. To overcome this problem, only 3 years data with minimum gaps were used for analysis. Short gaps were filled through interpolation.

2.6 Atmospheric Pressure

Atmospheric pressure is the pressure exerted by the atmosphere as a result of the Earth's gravitational pull upon a column of air above the measuring point. Generally, atmospheric pressure decreases with altitude. The World Meteorological Organization (WMO) accepts the mercury barometer as a primary standard for measurement of atmospheric pressure. Other types of probes include aneroid (vibrating diaphragm and aneroid capsule) and piezo (piezoelectric and piezoresistive).

Fig. 2.8 Three years of water temperature observations at three depths in Lake Okeechobee, south Florida

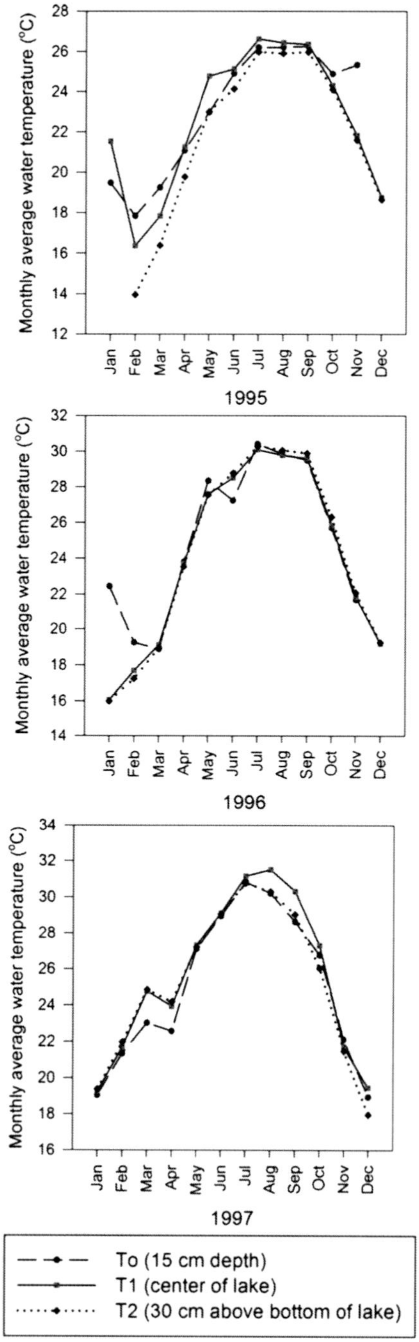

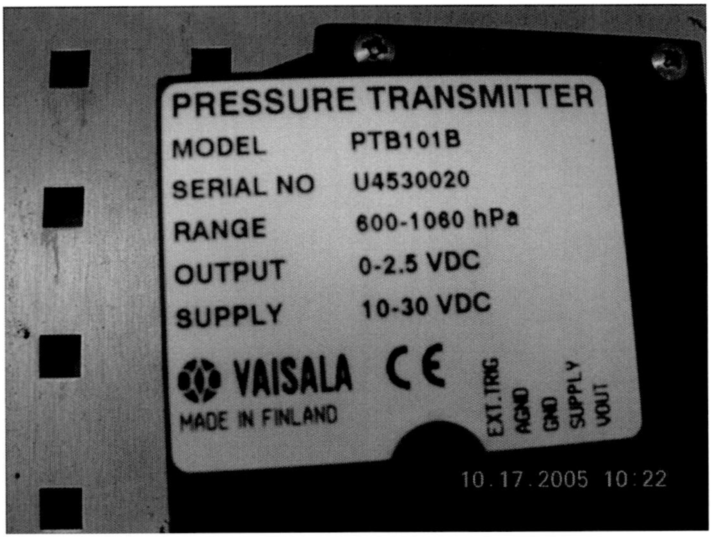

Fig. 2.9 PTB101B pressure sensor patented by Vaisala (Photograph provided by South Florida Water Management District)

Currently, digital atmospheric or barometer pressure measuring devices are pressure transducers with digital output. A type of a barometric pressure transducer is the PTB101B, fabricated from two pieces of silicon, with one piece acting as a pressure-sensitive diaphragm (Fig. 2.9). The range is 600–1,060 mb (hPa) with an accuracy of $\pm$ 0.45 mb. Units of atmospheric pressure are mm Hg, in Hg (inch mercury), mb (millibar), Pa (Pascal), and psi (pounds per square inch). A standard atmosphere is 760 mm Hg, 29.92 in Hg, 1013.25 mb, 101,325 Pa, and 14.696 psi.

The 2005 observations of atmospheric pressure over Lake Okeechobee, south Florida, from four weather stations inside the lake are depicted in Fig. 2.10. The surface area of the lake is 1,732 km^2. The mean atmospheric pressures at sites L001, L006, L005, and LZ40 were 102.69, 101.64, 101.59, and 101.53 kPa, respectively. It shows that the measurements are good quality. The lowest atmospheric pressure, 99.27 kPa, was on October 24, 2005, when Hurricane Wilma was passing through south Florida (Abtew and Iricanin 2008). Relatively, the atmospheric pressure data quality was very good with few missing, partial, or estimated data.

2.7 Wind Speed and Wind Direction

Wind speed is one of the required inputs in ET models that contain aerodynamic, mass, and momentum transfer components. The standard weather station has wind speed measurements at 10-m height. Generally, ET models require wind speed measurements at 2-m height. Based on the log profile of wind speed, estimates

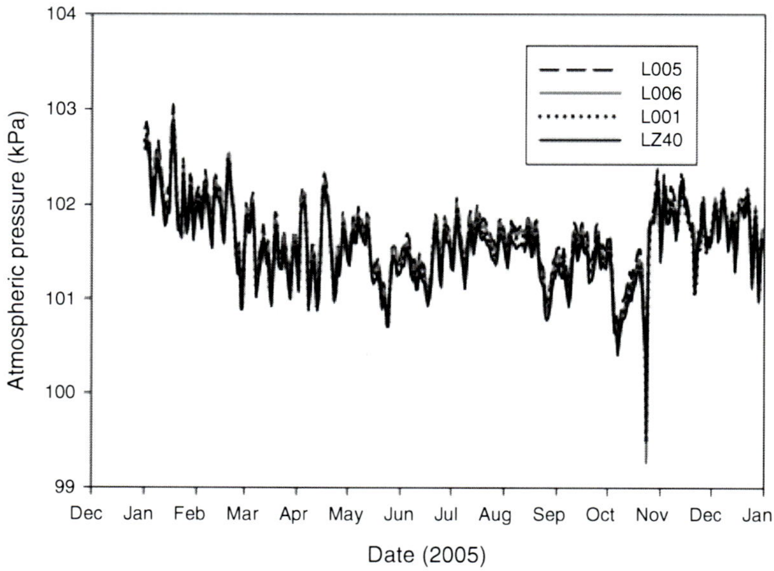

Fig. 2.10 Daily mean atmospheric pressure at four sites on Lake Okeechobee in south Florida with a hurricane event on October 24, 2005

Fig. 2.11 A Vaisala WS425 ultrasonic wind sensor at weather station L006 in Lake Okeechobee in south Florida (Photograph provided by South Florida Water Management District)

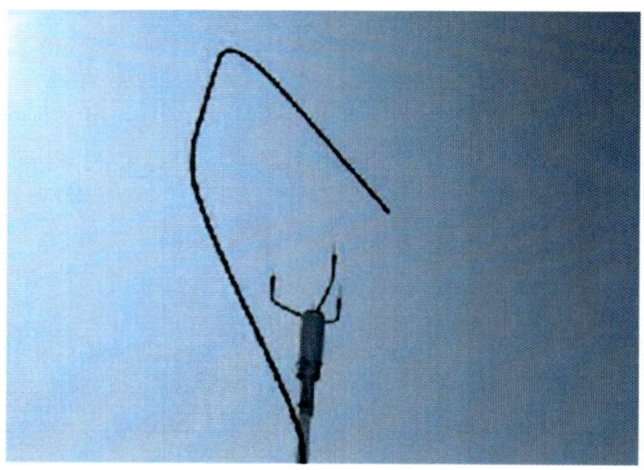

are generated using various equations. Cup and propeller type anemometers are common wind speed gauges. The sensors consist of a rotary part and a signal-generating part. The wind direction sensor is a wind vane that consists of a long airfoil-type tail which points to the direction of the wind. The vertical part of the wind vane drives a potentiometer which generates a signal proportional to the wind direction. Modern-type wind speed gauges are ultrasonic wind sensors such as the Vaisala WINDCAP WS425 (Fig. 2.11). According to the user's guide, the WS425 has an onboard microcontroller that captures and processes data and performs serial communications. The wind sensor has three equally spaced ultrasonic transducers on a horizontal plane with accuracy of measurement of $\pm$ 3% (Abtew et al. 2007).

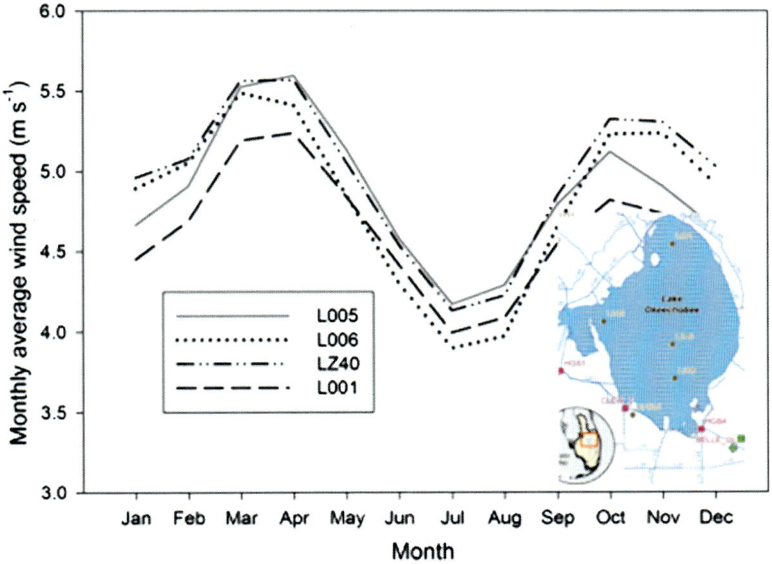

Fig. 2.12 Monthly mean wind speed at four sites in Lake Okeechobee, Florida (1994–2010)

Wind speed measurement sites require open spaces with no obstructions such as buildings, fences, and vegetation. Wind speed is a function of the characteristics of ground surface. The resistance to wind flow by ground cover such as vegetation results in determining the wind speed profile over that surface. Generally, wind speed data are relatively of good quality if sensors are installed at appropriate locations. But deficiency in gauge maintenance, calibration, and replacement can result in low-quality data collection or no data at all. As reported in Allen et al. (2005), failure in anemometers is manifested in registering constant small values (less than 0.5 $\mathrm{m\,s^{-1}}$ or the wind speed threshold for a new anemometer). Also, maximum and mean wind speed will have similar values matching numerical offset in the calibration equation. Incorrect readings may occur when large rainfall or ice pellets hit a transducer or when ice forms on anemometers. In the database used for illustrations in this chapter, there are missing, estimated, and partial observations. In general, there is sufficient good quality data to characterize daily wind speed over the lake where weather stations are located. Gaps in a site data set were filled with average values from other nearby sites that had observations for that day. Comparison of monthly mean wind speed at the four lake sites is presented in Fig. 2.12.

The mean wind speeds for sites L005, L006, LZ40, and L001 are 4.86, 4.83, 4.97, and 4.62 $\mathrm{m\,s^{-1}}$, respectively. But also, there are periods where some sites demonstrated errors. The following short-period illustration shows systematic error in wind speed measurements at two of the three sites in Lake Okeechobee, south Florida (Fig. 2.13a).

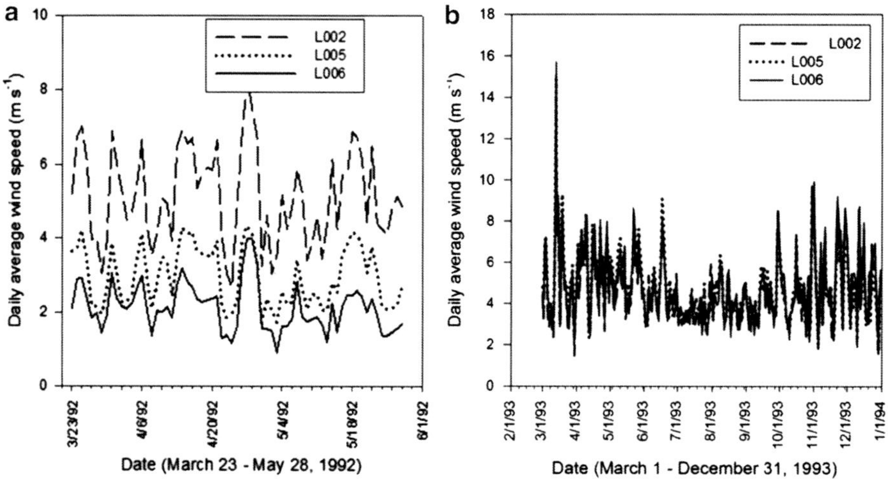

Fig. 2.13 Wind speed measurements at three sites in Lake Okeechobee, (**a**) systematic bias and (**b**) observations without bias

The persistent low readings at site L005 and L006 are biased with systematic shift to lower readings compared to mean values for March through May (Fig. 2.12). Figure 2.13b depicts observations without any bias for the same three sites.

2.7.1 *Wind Profile*

Wind profile, the change of wind speed with height, fits a logarithmic function. Wind speed increases with height. When wind speeds are not measured at desired heights, estimates can be derived from data measured at different heights using the logarithmic equation (Eq. 2.3):

$$u_{z_2} = u_{z_1} \frac{\ln \frac{z_2 - d}{z_o}}{\ln \frac{z_1 - d}{z_o}} \tag{2.3}$$

where u_{z_2} is wind speed at the desired height of z_2, u_{z_1} is wind speed at the measurement height of z_1, d is displacement height, and z_o is roughness height. Measured wind speed variation with height is depicted in Fig. 2.14.

There are several equations to estimate displacement height (d) and aerodynamic roughness (z_o), as shown in Abtew et al. (1989). Applying the author's methods (Eqs. 2.4 and 2.5), wind speed is estimated at 2 m (Abtew et al. 1989):

$$d = F_c h_c \tag{2.4}$$

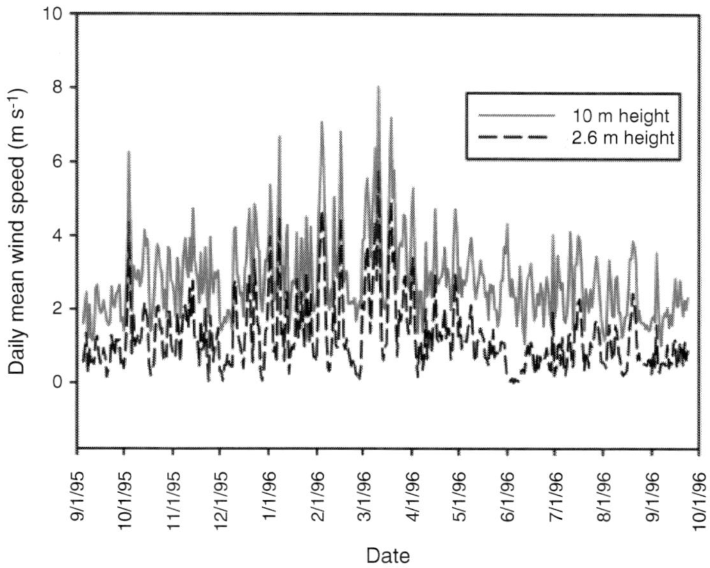

Fig. 2.14 Wind speed measurements at two heights in south Florida in the Everglades Nutrient Removal Project, a constructed wetland

where F_c is fraction of surface cover and h_c is average height of cover.

$$z_o = 0.13\,(h_c - d) \tag{2.5}$$

At the site of wind speed measurement, the estimated fraction of wetland vegetation cover was 0.8 with average height of 1.5 m (Abtew and Obeysekera 1995). From Eqs. 2.4 and 2.5, d is 1.20 m and z_o is 0.039 m. In cases where wind speed is measured at two heights but data need is at a different height (2 m), in Eq. 2.1, the weighted average of the two wind speeds and heights can be used with more weight given to wind speed and height of measurement closer to the height of interest. The weights are based on the proportion of distance of wind speed measurement height and the desired height. As an illustration, 2-m wind speed is estimated from 2.6 to 10-m wind speed measurements shown in Fig. 2.14. The weights are 13.33 for height 2.6 m and 1 for height 10 m. The result is shown in Fig. 2.15a, b. The logarithmic fit of wind profile for mean wind speed at the three different heights is shown in Fig. 2.15b.

2.7.2 *Wind Barrier's Impact on Wind Speed and Pattern*

Wind barriers such as fences, vegetation, and buildings affect wind speed measurements. Decrease in wind speed on the leeward side of barriers (windbreaks,

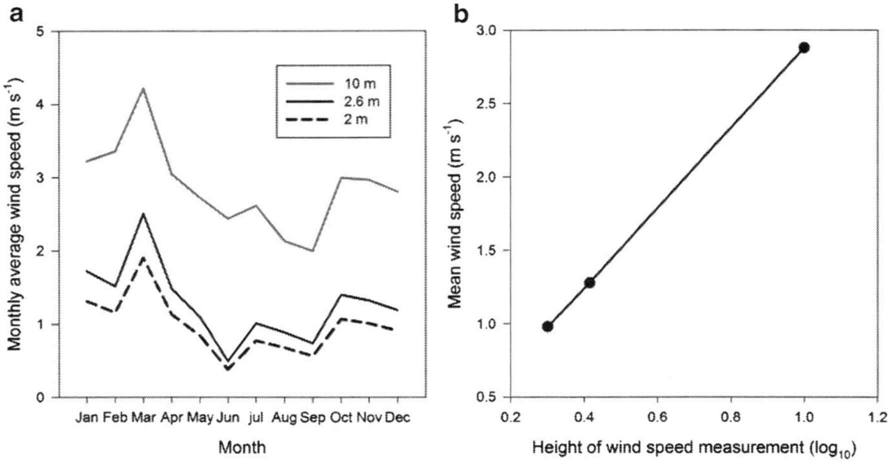

Fig. 2.15 (**a**) Monthly average wind speed at two measurement heights (2.6 m, 10 m) and estimates at 2 m height and (**b**) logarithmic relationship of wind speed and height of measurement

shelterbelts, snow fences, etc.) has been reported through wind tunnel and field studies. As obstruction to airflow, barriers bring about three effects on the surrounding environment. The flow of the approaching wind is changed in magnitude and direction when it crosses the barrier. The leeward airflow pattern is changed. The leeward microenvironment, temperature, vapor pressure, and evapotranspiration are altered due to the barrier. The impact of wind barriers extends to a leeward length of 30 times the height of the barrier. Porous barriers like fences have leeward impact exponentially related to percent porosity. Figure 2.16 depicts the relationship of the ratio of leeward to windward wind speed (U_L/U_W), percent barrier porosity, and the ratio of leeward distance from barrier in barrier heights (X) for distances 5–30 times the barrier height (Borrelli et al. 1989). The region behind the barrier downwind up to a distance of five times the barrier height ($X = 5$) is the vortex area.

2.8 Solar Radiation

The principal source of heat energy for the planet is solar radiation. The amount of solar energy received at a location is dependent on time of the day, day of the year, latitude, altitude, and cloud cover. The amount of net solar radiation received is further dependent on the reflectance of the receiving surface. The solar flux comes in the range of 0.1–3.2 μm wavelength with the visible range from 0.4 to 0.7 μm. Large energy flux comes from the Sun, the extraterrestrial solar flux (R_A), and what passes through the atmosphere and reaches the Earth's surface is a fraction of R_A. The maximum solar radiation that reaches the Earth's surface

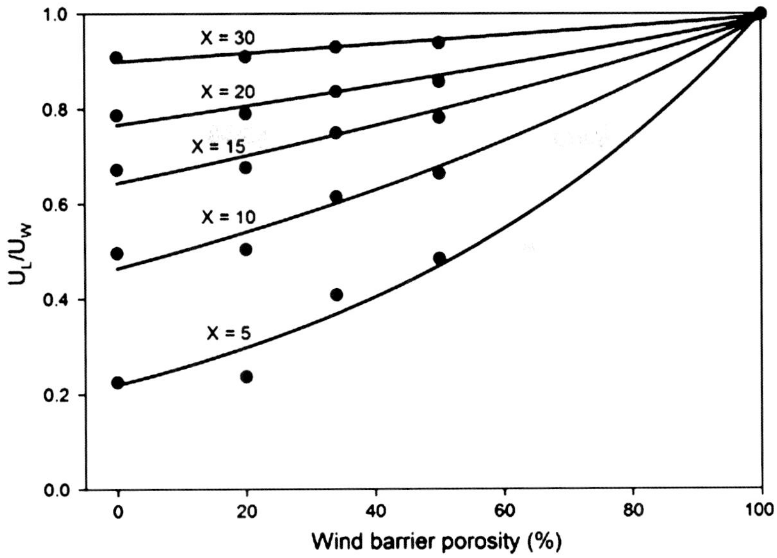

Fig. 2.16 Wind barrier porosity, height, leeward distance, and wind speed relationship

at a given time and location is when conditions of clear sky occur (R_{so}). Since cloud cover reduces incoming solar radiation (R_s), R_s is lower than R_{so}. There have been developed various equations to estimate R_{so} from R_A and R_s from R_{so}. R_A is also estimated from variables and constants that determine the solar flux the Earth receives. Equations 2.6, 2.7, 2.8, and 2.9 are few of the equations in the literature (Allen et al. 2005):

$$R_A = \frac{24}{\pi} G_{sc} d_r \left(\omega_s \sin(\varphi)\sin(s\delta) + \cos(\varphi)\cos(s\delta)\sin(\omega_s)\right) \tag{2.6}$$

where R_A is extraterrestrial radiation (MJ m^{-2} day^{-1}), G_{sc} is solar constant (4.92 MJ m^{-2} h^{-1}), d_r is inverse relative distance factor (squared) for the Earth–Sun (non-dimensional), ω_s is sunset hour angle (radians), φ is latitude (radians), and $s\delta$ is solar declination (radians).

$$R_{so} = \left(0.75 + 2 \times 10^{-5}\text{elev}\right) R_A \tag{2.7}$$

where R_{so} is clear sky solar radiation (MJ m^{-2} day^{-1}), R_A is extraterrestrial radiation (MJ m^{-2} day^{-1}) , and elev is site elevation above sea level (m). For south Florida and other parts of the world where elevations are closer to sea level, Eq. 2.7 can be simplified as follows:

$$R_{so} = 0.75\, R_A \tag{2.8}$$

Fig. 2.17 A LI-COR LI-200S pyranometer (Photograph provided by South Florida Water Management District)

Incoming solar radiation is a function of cloud cover (sunshine hours, s). An estimation equation of incoming solar radiation (R_s) from clear sky solar radiation (R_{so}) is given as follows (Jensen 1974):

$$R_s = (0.35 + 0.61\,s)\,R_{so} \tag{2.9}$$

where R_s and R_{so} have the same unit.

Incoming solar radiation (R_s) is measured as energy flux density of both direct beam and diffuse sky radiation passing through a horizontal plane of known area (1 m^2). Solar radiation is measured with a pyranometer such as the LYCOR Model LI-200S (Fig. 2.17). Solar radiation varies diurnally (Fig. 2.3), seasonally, by atmospheric conditions, and by location. The amount of clear-day solar radiation that reaches different locations is available from published tables or can be computed using Eq. 2.7. Since atmospheric conditions such as cloud cover reduce the amount of radiation that reaches the Earth, either cloud cover or radiation requires monitoring. In places where pyranometer instrumentation is not available, hours (percent) cloud cover is recorded, and solar radiation is derived as a percentage of clear-day radiation.

Calibration and maintenance of the pyranometer are required to collect good quality data. The LYCOR Model LI-200S pyranometer is calibrated against an Eppley precision pyranometer of which the calibration is periodically confirmed. The uncertainty of the calibration is $\pm 5\%$ (Kinsman et al. 1994). Major maintenance work is keeping the sensor clean, free from obstruction and routine calibration. Pyranometers are durable and relatively easier to acquire good quality data compared to net solar radiation sensors and radiometers. Data is reported in kW m^{-2} with a range of 0–1. The daily solar radiation is integration of the total energy received for the day in MJ m^{-2} day^{-1}. Units of solar radiation include Calorie cm^{-2} day^{-1}, Langley day^{-1}, MJ m^{-2} day^{-1}, and mm water day^{-1}. The water depth unit comes from latent heat of vaporization of water. Conversion of these units is shown in Table 2.2.

Good quality solar radiation data collection requires a trained technician and good calibration and maintenance program. Otherwise, quality of data will be low.

Table 2.2 Solar radiation energy flux and energy conversion units

Unit	Equivalent unit
1 cal cm^{-2} day^{-1}	0.041868 MJ m^{-2} day^{-1}
23.884 cal cm^{-2} day^{-1}	1 MJ m^{-2} day^{-1}
1 cal cm^{-2}	I Langley
1 Langley	4.1855 J cm^{-2}
1 cal cm^{-2}	3.6855 BTU ft^{-2}
1 cal cm^{-2}	0.069758 W cm^{-2}
1 cal cm^{-2}	0.69758 kW m^{-2}
1 MJ m^{-2} day^{-1}	0.408 mm of water day^{-1}
1 mm of water day^{-1}	2.45 MJ m^{-2} day^{-1}
1 mm of water day^{-1}	58.6 cal cm^{-2} day^{-1}
1 MJ m^{-2} day^{-1}	86.4 kW m^{-2}
1 W m^{-2}	0.0864 MJ m^{-2} day^{-1}
1 W m^{-2}	2.064 cal cm^{-2} day^{-1}

Common data quality issues include missing observations, partial observations, and erroneous observations. The clear sky solar radiation is the maximum limit to incoming solar radiation. For every area, solar range and mean values can be known from observations, and information can be used for data quality evaluation. As an illustration of data quality, Fig. 2.18 depicts the 1995 daily solar radiation from four weather stations on Lake Okeechobee, south Florida.

Displayed also are the extraterrestrial solar flux and clear sky solar radiation. Incoming solar radiation data greater than $R_{so}$are erroneous. Sensor problems at site L005 are apparent with positive bias in the first half of the year and negative bias in the later part of the year. It should be possible to evaluate daily data from all the sites and develop a composite improved data set for application. Figure 2.19 depicts daily solar radiation measurements in Belle Glade, south Florida, from 1992 to 2009. It is apparent there is positive bias in data for the first 4 years.

2.9 Net Solar Radiation

Net solar radiation (R_n) is net shortwave radiation, which is the balance from incoming solar radiation (R_s) and reflected back solar radiation. The Earth's surface reflects back a portion of the incoming solar radiation. The reflection fraction depends on the characteristics of surface cover and is described by the term albedo (α), as shown in Eq. 2.10. Water albedo is latitude-dependent (Cogley 1979). Albedo also changes with time of day and time of season due to change in Sun angle, but usually a single value is used at all times (Allen et al. 2005):

$$R_n = (1 - \alpha)\, R_s \tag{2.10}$$

Net solar radiation is measured with radiometers. Radiometers are susceptible to environmental fouling, and daily inspection is recommended. Accuracy of the

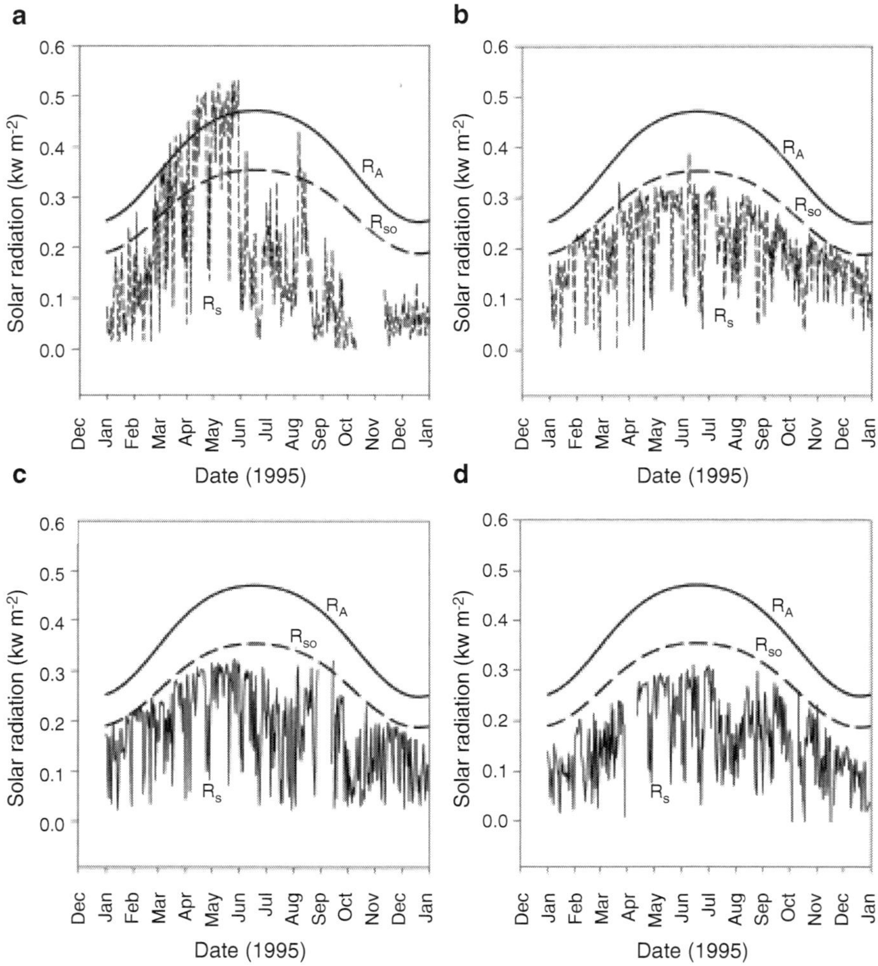

Fig. 2.18 Daily measured R_s at (**a**) L005, (**b**) L006, (**c**) L001, and (**d**) LZ40 sites in Lake Okeechobee, south Florida, and computed R_A(Eq. 2.6) and R_{so} (Eq. 2.7)

hemisphere is degraded by birds, dust, and other fouling. For best performance, daily inspection and maintenance are recommended (Kinsman et al. 1994). Routine check on calibration is a must to acquire good quality data. At each inspection, the outer surface of the hemisphere should be cleaned with dry lint-free cloth. Moisture should be removed from the internal surface of the hemisphere.

Maintenance should include checking and replacing desiccators. The Q7.1 is a type of radiometer (Fig. 2.20). It is a high-output thermopile sensor that measures the algebraic sum of incoming and outgoing short- and longwave radiation. Incoming radiation consists of direct (beam) and diffuse solar radiation plus longwave irradiance from the sky. Outgoing radiation consists of reflected solar radiation

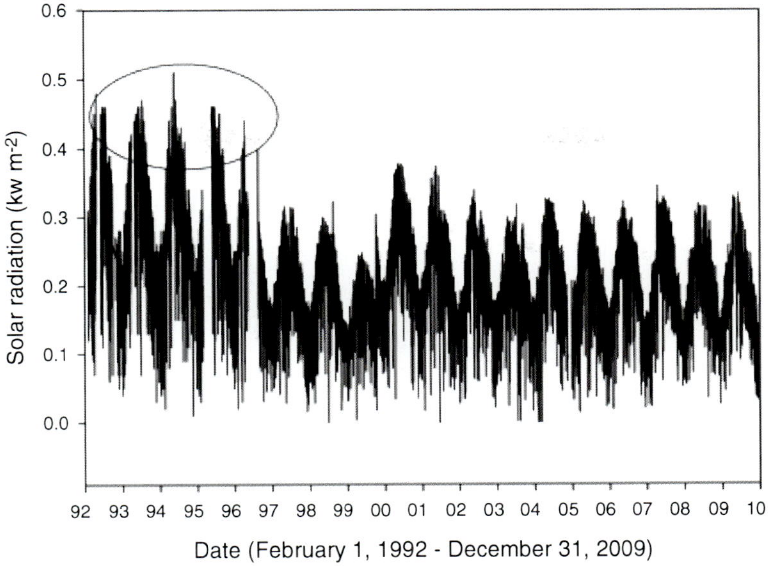

Fig. 2.19 Daily solar radiation observations in Belle Glade, south Florida, with positive bias in data

Fig. 2.20 A Q7.1 net radiometer from a weather station in south Florida (Photograph provided by South Florida Water Management District)

plus the terrestrial longwave component (Abtew et al. 2007). Probably, the most problematic sensor is the radiometer to collect continuous good-quality data. Due to this problem, evaporation and ET estimation models that require net solar radiation incur error of estimation. Data evaluated in this analysis has missing data, partial observations, and estimates. An illustration of net solar radiation observations with biases, from three weather stations in south Florida, is shown in Fig. 2.21a.

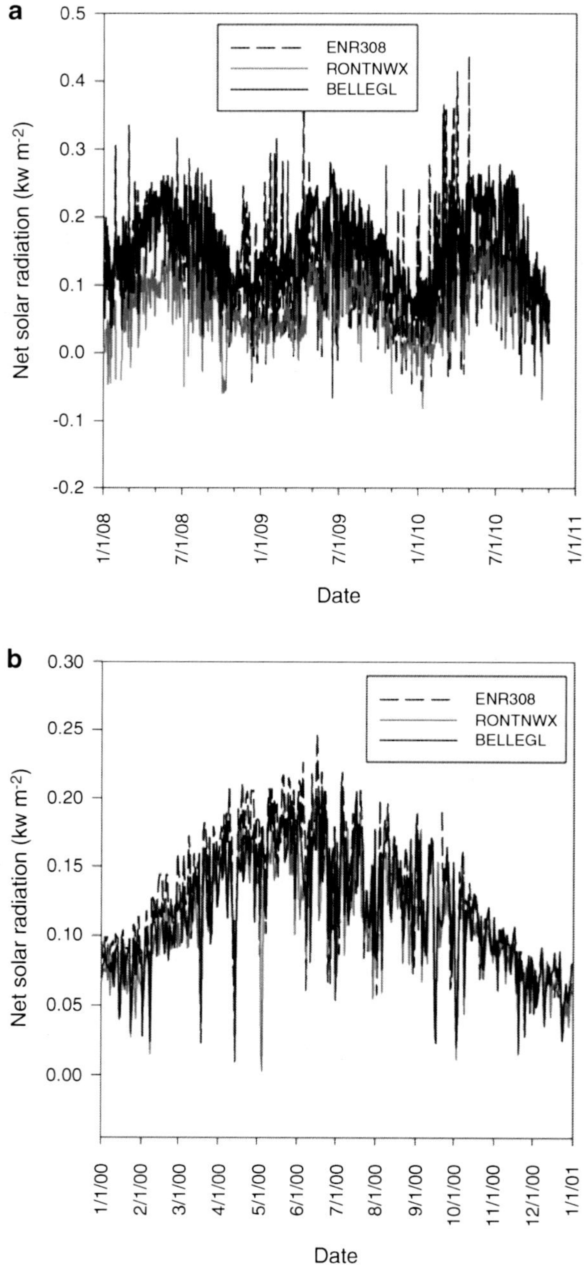

Fig. 2.21 Net solar radiation measurements at three sites in south Florida; (**a**) data with bias and (**b**) good quality data

The mean net solar radiation was 0.12, 0.07, and 0.15 kW m^{-2} for south Florida sites ENR308, ROTNWX, and BELLEGL, respectively. Figure 2.21b shows better quality net radiation data from the same three sites for a different time period. The mean net solar radiation was 0.13, 0.11, and 0.12 kW m^{-2}, respectively.

2.10 Summary

Land-based meteorological monitoring is an expensive undertaking. Acquiring good quality data increases the cost of operation and maintenance. Historically, meteorological data has been collected by different instrumentation, technician skill, and maintenance frequency and quality. Data quality of a monitoring network reflects the effort put in data collection by the monitoring system. Unfortunately, data users have to screen and evaluate the quality of data available from different sources. In most cases, missing data estimation and combining data from different sources are required to get acceptable quality data for evaporation and evapotranspiration estimation. Quite often, meteorological data is applied for ET estimation without scrutiny of data quality and becomes a major source of ET estimation error.

Acknowledgements We would like to acknowledge Dr. Qinglong (Gary) Wu from South Florida Water Management District for taking the weather station and sensor photographs.

References

Abtew W, Iricanin N (2008) Hurricane effects on South Florida Water Management System: a case study of Hurricane Wilma of October 2005. J Spat Hydrol 8(1):1–21

Abtew W, Obeysekera J (1995) Lysimeter study of evapotranspiration of cattails and comparison of three estimation methods. Trans ASAE 38(1):121–129

Abtew W, Borrelli J, Gregory JM (1989) Wind profile: estimation of displacement height and aerodynamic roughness. Trans ASAE 32(2):521–527

Abtew W, Pathak C, Huebner RS, Ciuca V (2007) Chapter 2, Hydrology of the South Florida environment. Appendix 2-2: Hydrologic monitoring network of the South Florida Water Management District. In: Pathak C (ed) 2007 South Florida environmental report. South Florida Water Management District, West Palm Beach, FL. www.sfwmd.gov/sfer. Accessed 3 Aug 2012

Allen RG, Walter IA, Elliott R, Howell T, Itenfsu D, Jensen M (eds) (2005) The ASCE standard reference evapotranspiration equation. ASCE, Reston

Borrelli J, Gregory JM, Abtew W (1989) Wind barriers: a reevaluation of height, spacing, and porosity. Trans ASAE 32(6):2023–2027

Bosen JF (1958) An approximation formula to compute relative humidity from dry bulb and dew point temperatures. Mon Weather Rev 86(12):486

Cogley JG (1979) The albedo of water as a function of latitude. Am Meteorol Soc 1979(June): 775–781

Crowell ML, Mtundu ND (2000) Guidelines for quality control and quality assurance of hydrologic and meteorological data. Volume 2: Data management. South Florida Water Management District, West Palm Beach

Jensen ME (ed) (1974) Consumptive use of water and irrigation water requirements. Report prepared by Technical Committee on irrigation water requirements. Irrigation and Drainage Division, ASCE, New York

Jin K, James RT, Lung W, Louks DP, Park RA (1998) Assessing Lake Okeechobee eutrophication with water quality models. J Water Resour Plan Manage 124(1):22–30

Kinsman G, Kite J, Mtundu D (1994) Guidelines for the collection of hydrologic and meteorological data volume I: field applications. South Florida Water Management District, West Palm Beach

World Meteorological Organization (1996) Guide to meteorological instruments and methods of observation. WMO-No. 8. Secretariat of WMO, Geneva

Chapter 3
Evaporation and Evapotranspiration Measurement

Abstract Direct measurements of evaporation and evapotranspiration (ET) are usually experimental work where the data is used to calibrate models for long-term estimation from meteorological variables. Even the evaporation pan that is commonly used requires calibration coefficients to relate to the local evaporation or evapotranspiration. Advancement in evaporation and ET measurements will always improve the performance of ET estimation models. In this chapter, measurement methods such as pan, lysimeters, water budgets, and advanced sensor application are presented in detail.

Keywords Evaporation • Evapotranspiration • Pan evaporation • Lysimeter • Eddy correlation • Bowen ratio • Lidar

3.1 Introduction

Evaporation from open water and ET from vegetated surfaces are critical parameters of hydrology such that efforts to measure and estimate these parameters are justified. Advances in measurements and estimation have followed advances in technology. The progress of measurement techniques ranges from the evaporation pan to remote sensing techniques. In this chapter, the methods of evaporation and ET measurements are presented. Actual applications of the methods are discussed in this and other chapters in this book.

3.2 Pan Evaporation

The evaporation pan is the most common and probably the oldest widely used method of measurement or estimation of open water evaporation. There are various types of pans that are used in different parts of the world. A common pan is the

W. Abtew and A. Melesse, *Evaporation and Evapotranspiration: Measurements and Estimations*, DOI 10.1007/978-94-007-4737-1_3,

Fig. 3.1 Variation in site environment and setup for four evaporation pans (Abtew et al. 2011)

National Weather Service Class A evaporation pan. The pan is 120.7 cm in diameter and 25 cm in depth. Water is added or removed to maintain water level at 5 cm from the rim. The sunken Colorado pan is square in shape (100 cm × 100 cm), 50 cm deep, and buried in the ground to a depth of 45 cm. Water level is monitored with a hook gauge and stilling well. Measurement resolution is 1 mm and accuracy is ± 1 mm (Crowell and Mtundu 2000). The pan is accompanied with a rain gauge to factor out the contribution of rainfall to the depth of water in the pan. In some cases, a partial- or full-scale weather station may accompany pan evaporation stations.

Variations between pans include setup and the pan environment (Fig. 3.1). Pan setups vary from an elevated stand (Fig. 3.1), on a platform on the ground and fenced (Fig. 3.1), behind a structure (Fig. 3.1) and with bird guard (Abtew et al. 2011).

Daily evaporation from the evaporation pan is derived from a mass balance equation, Eq. 3.1:

$$E_{\text{pan}} = D_{t-1} - D_t + R_{\text{f}} - L \pm e \tag{3.1}$$

where D_t is current day depth of water in the pan and D_{t-1} is previous day depth of water, measured from the top; R_{f} is rainfall over the pan; L is other losses such as bird or animal consumption; and e is errors.

Sources of error in monitoring evaporation with an open outdoor pan include environmental factors such as location, installation, wind flow obstruction, advective heat sources or losses in the area surrounding the pan, height of pan, bird guard, rate of windblown sediment accumulation, algae growth in pan, and frequency of cleanup. Bird guard was acknowledged for lowering evaporation rates. In an Australian case, a correction factor (7%) has been applied to correct for the effect of bird guard (Gifford et al. 2007). The accuracy of change in water level measurement through water level reading or measurement of volume of replacement water to fill the pan to previous day level is also a major source of error. The training and discipline of pan evaporation operators and pan maintenance frequency probably account for significant variations in pan evaporation data in close locations. Measurement of rainfall contribution to the pan is another potential source of error (Abtew et al. 2011). Rainfall splash in or out of the pan could occur. Observation recording, data transmission, and data storage are also potential processes where errors can occur.

Open water evaporation (E_{o}) is estimated from pan evaporation (K_{pan}) using the pan coefficient, K_{p} (Eq. 3.2). Reference crop evapotranspiration (E_{tp}) is estimated from pan evaporation using Eq. 3.3. In general, pan evaporation is higher than open water evaporation and reference crop ET, and the coefficients K_{p} and C_{et} are lower than 1. Pan coefficient, K_{p}, is dependent on the type of pan environment and operations (Abtew 2001). Reference crop coefficient, C_{et}, is dependent on meteorological conditions as wind speed and humidity and the values range from 0.35 to 0.85 (Jensen 1974):

$$E_{\text{o}} = K_{\text{p}}\, E_{\text{pan}} \tag{3.2}$$

$$E_{\text{tp}} = C_{\text{et}}\, E_{\text{pan}} \tag{3.3}$$

In evaluating historical records of pan data, factors such as relocation of the pan, changes in measuring gauges, and changes in operators have to be considered as factors influencing observations of evaporation. Historical pan evaporation data are usually plagued with outliers, gaps, and data of questionable quality. Variations in pan evaporation data within relatively small distances indicate the challenges of acquiring consistent observations from pans, as shown in Chap. 6. Comparison of pan evaporation data from seven sites around Lake Okeechobee in south Florida resulted in pan coefficients ranging from 0.64 to 0.95 demonstrating that each pan is influenced by local environment and operations (Abtew 2001). The wide range of pan coefficients tends to overemphasize the shortcoming of pan data (Shuttleworth 1993).

3.3 Lysimeters

A field lysimeter is a way of controlling a small section of the surrounding environment for water balance monitoring with little alteration to the physical and climatic conditions that prevail at the site. Evaporation or ET is derived from water balance. In most cases, the main part is a tank where soil is filled and vegetation is planted and devices to control and measure change in moisture are installed. There are two types of lysimeters: the weighing lysimeter and the water balance lysimeter. Historically, lysimeters have been used to measure ET from agricultural crops and use the data to calibrate ET estimation equations or models. The weighing lysimeter requires dry conditions under the crop for setup and monitoring. The water balance lysimeter is adaptable to a wetland environment. Due to cost of maintenance and operation, lysimeters are not used for monitoring ET continuously.

3.3.1 Weighing Lysimeter

A weighing lysimeter is a setup to measure ET as a difference in weight of the lysimeter. A hole is dug in the ground, and a stable concrete platform is built at the bottom where weight sensors (load cell) rest beneath a large tank filled with soil and vegetation from the area. The rim of the tank is flush with the surrounding area to simulate conditions at the site of installation, usually a farm field. Weighing lysimeters are expensive to install and operate.

3.3.2 Water Balance Lysimeter

A water balance lysimeter can be designed, installed, and operated under saturated and ponding conditions simulating a wetland environment. Different instrumentation is required when the lysimeter is operating under unsaturated conditions or where the water table is below ground.

3.3.2.1 Wet Lysimeter

A wet lysimeter is a lysimeter designed to measure ET under wet conditions where the water table is above ground. These types of lysimeters are easier to operate as input and output of water and water depth change can be relatively measured with ease. ET derivation from a ponding or wetland lysimeter is expressed as follows (Eq. 3.4):

$$\mathrm{ET} = R_{\mathrm{f}} + I - O + \delta \tag{3.4}$$

where R_{f} is rainfall, I is depth of water added, O is depth of water removed, and δ is change in depth of water in the lysimeter computed as follows (Eq. 3.5):

$$\delta = d_t - d_{t-1} \tag{3.5}$$

where d_t is depth of water level in lysimeter at time t and d_{t-1} is depth of water in lysimeter at time $t-1$. Depth is measured from the bottom.

3.3.2.2 Design of a Wet Lysimeter

Three wet lysimeters were installed in cattail, mixed marsh, and open water wetlands to measure wetland vegetation ET and open water evaporation at the Everglades Nutrient Removal Project constructed wetland in south Florida (Abtew and Hardee 1993; Abtew and Obeysekera 1995; Abtew 1996). The main feature was a polyethylene tank of 3.53 m diameter and 91.44 cm depth. The tank was placed on a square steel frame. Anchors at each corner were used to level the tank and place the rim of the tank at the preferred elevation. The objective was to place the rim of the tank close to water level in the marsh for most of the time. This was to maintain similar conditions in the lysimeter and the wetland.

A network of 18 m of perforated 5 cm diameter PVC pipe was laid in the bottom and along two sides. The pipes extending out are for inflow and outflow pumping (Fig. 3.2a). The perforated pipes are wrapped with filter cloth to minimize pump clogging during pumping in and out. The advantage of the pipe network is to quickly equalize water level in the tank during pumping in and out. On the opposite side, there is a stilling well for water level monitoring (Fig. 3.2b). The stilling well is also perforated and covered with filter cloth to stabilize water level quickly.

The inflow and outflow pumps are two-way compact self-priming marine utility pumps, TEEL Water Systems Model IP580E, 12 V DC, with a unit pump discharge capacity of 968 L/h at 1.52 m total head. Both pumps were controlled by a Campbell Scientific, Inc. Model CR10 data logger programmed to turn an electric switch on and off when the desired water level is reached. Flow rate was measured by Data Industrial 4000 series Model 402200 flow meters. The power source for the CR10 data logger was a 12-V battery (8-Ah rating) recharged with a 5.2-W solar panel. The power source for the pumps was a 12-V marine battery (80-Ah rating) recharged with an 18-W solar panel.

Water level in the lysimeter was measured with an analog evaporation gauge, Model 6844-A. The output range of the gauge was 0–14.22 cm with a measuring accuracy of ±0.38 mm of true water level. The data logger records depth to water

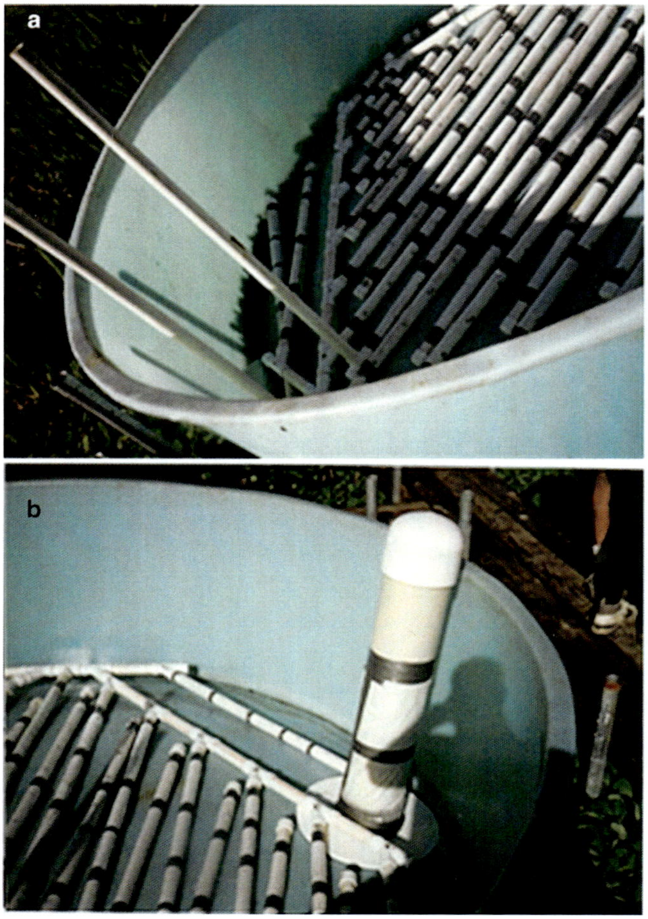

Fig. 3.2 Network of pipes inside a wet lysimeter, (**a**) inflow and outflow pipes and (**b**) stilling well (Photograph provided by South Florida Water Management District)

level, inflow, outflow, and time at which pumps were turned on and turned off. The lysimeter was a fully automated system where data were scanned every 5 s and registered at a 15-min interval.

Soil and vegetation was filled from the surrounding area, and in a short time, the lysimeter looked like the surrounding wetland. Figure 3.3a–c depicts stages from soil filling, instrumentation, and planting to fully operational state for a cattail lysimeter. The lysimeter was accompanied by a weather station with solar radiation, net solar radiation, photosynthetic radiation, humidity, air temperature, and atmospheric pressure measurement at 2-m height. Wind speed and direction was measured at 10 m height. A second anemometer was installed at 2-m height. Data was scanned every 5 s and registered at 15-min intervals. Meteorological data from the weather station was used to calibrate ET models using the lysimeter data as a reference.

Fig. 3.3 (**a**) Soil filling, (**b**) instrumentation and planting, and (**c**) fully operational lysimeter with cattails (Photograph provided by South Florida Water Management District)

The lysimeter was operated within a range of 3.81 cm of water level difference. Water was pumped out when the level was 3.81 cm from the top until the level reached 6.35 cm. Water was pumped in when the level is 7.62 cm from the top and filled to 5.08 cm from the top. Water was pumped in from the marsh through a pipe where the section of the pipe in the marsh is perforated and wrapped with filter cloth. The outflow pump discharged into the marsh. Test and calibration of the flow meters and the analog evaporation gauge was performed in a lab before installation.

Fig. 3.4 (**a**) Cattails, (**b**) mixed marsh, and (**c**) open water lysimeters (Photograph provided by South Florida Water Management District)

Three fully operational lysimeters in cattail marsh, mixed vegetation marsh, and open water marsh are shown in Fig. 3.4a–c.

3.3.2.3 Design of a Dry Lysimeter

A dry lysimeter was installed in the Everglades Nutrient Removal Project constructed wetland in south Florida to measure ET under saturated and unsaturated conditions (Abtew et al. 1998). The difference between the wet and dry lysimeters is that the wet lysimeter was operated as a wetland with ponding water above the soil surface. The dry lysimeter water table was below the soil surface, and it featured saturated and unsaturated soil profiles. The tank and setup was similar to the wet lysimeter. The unsaturated moisture content was measured indirectly with a combined electrical soil moisture and temperature sensor (AQUA-TEL, Model 29+T, Automata Inc., Grass Valley, CA). The 74 cm long sensors measure the dielectric constant of the soil. The dielectric constant is directly related to soil moisture content. The sensor averages moisture content of the soil and soil temperature through the soil profile. The change in water content of the saturated zone can be computed from the change in water level in the saturated zone and the soil water-holding capacity. The water table in the soil was monitored with

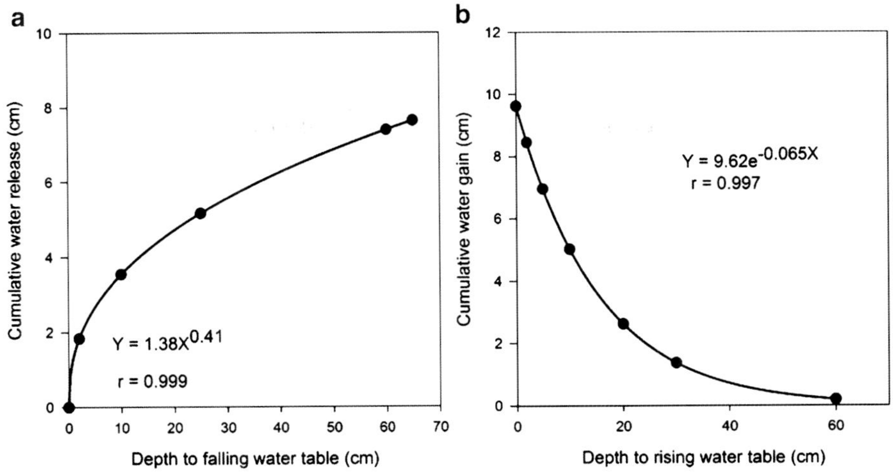

Fig. 3.5 Soil water-holding capacity curves under a (**a**) falling and (**b**) rising water table (Abtew et al. 1998)

redundant water level gauges, an SDI float mechanism and a pressure transducer. Figure 3.5 shows test results of the soil water-holding capacity under falling (a) and rising (b) water table conditions (Abtew et al. 1998).

A pump-in and pump-out test was run to determine the soil water storage capacity under rising and falling water table conditions. The soil moisture output range was 0–1 mA. The output was changed to mV (0–2,500) using a 2.5-kΩ shunting resistor. The soil type was not given in the calibration curves for the moisture sensors; gravimetric soil moisture content analysis was required to develop a calibration curve for the soil moisture sensors. Evapotranspiration can be estimated from change in soil moisture storage (SM), rainfall (R), water added (I), and water removed (O) as follows (Eq. 3.6):

$$\mathrm{ET} = \Delta\mathrm{SM} + R + I - O \tag{3.6}$$

3.4 Eddy Correlation

Eddy correlation is the covariance between two variables associated with turbulent wind motion. The method is based on the correlation of vertical wind speed and air moisture content fluctuation. At the surface, wind speed is parallel to the ground, but there are eddies with a positive or negative vertical wind speed (w') component at an instant but with mean of zero vertical wind speed. The air mass with vertical wind speed has specific humidity (q'), a fluctuation from the mean specific humidity of the air ($\bar{q}$). When positive w' and positive q' coincide, then moist than normal air

Fig. 3.6 Eddy correlation instrumentation in the Everglades (German 2000; U.S. Geological Survey)

is carried away from the ground surface. When negative w' and negative q' coincide, drier than normal air moves toward the ground (Shuttleworth 1993). Vapor flux (E) is computed by Eq. 3.7:

$$E = \overline{w'q'} = \frac{1}{N}\sum_{i=1}^{N}(w_i - \bar{w})(q_i - \bar{q}) \tag{3.7}$$

where w_i is vertical wind speed and q_i is specific humidity at time i. The eddy correlation instrumentation requires highly maintained fast responding sensors. Daily maintenance is required, and there is no guarantee of collecting continuous good quality data. Figure 3.6 depicts eddy correlation instrumentation.

3.5 Bowen Ratio

Estimation of sensible heat, H, in the energy balance is challenging, as shown in Chaps. 4 and 8. Temperature change with height and a transfer coefficient is required to estimate H. The Bowen ratio method substitutes the Bowen ratio (β) in the energy balance equation in place of H (Eq. 3.8). The Bowen ratio is the ratio of sensible heat to latent heat flux (Eq. 3.9):

$$\lambda E = \frac{R_{\mathrm{n}} - G}{1 + \beta} \tag{3.8}$$

Fig. 3.7 Bowen ratio instrumentation site in the Everglades (German 2000; U.S. Geological Survey; Abtew 2005)

where λE is latent heat flux, R_n is net solar radiation, and G is soil heat flux.

$$\beta = \frac{H}{\lambda E} = \gamma \frac{\Delta T}{\Delta e} \tag{3.9}$$

where γ is psychrometer constant, ΔT is change in temperature, and Δe is change in vapor pressure.

The Bowen ratio estimation requires temperature and vapor pressure measurements at two heights over the water surface. Figure 3.7 depicts a Bowen ratio instrumentation in the Everglades measuring temperature and humidity at two heights to determine T and e. These sensors are placed on a movable mechanism where the lower and upper sensors exchange position every 15 min to minimize instrument bias (German 2000). In Fig. 3.7, sensors are marked, wind speed and direction (1), stilling well for water level measurement (2), pyranometer (3), rain gauge (4), air temperature and humidity sensors at two heights (5), net radiometers (6), data logger and phone (7), and solar panel (8).

Different approaches have been presented to avoid temperature measurements at two heights. As a substitute, water surface temperature at the bottom and air temperature above the water are used to estimate β with associated saturation, actual vapor pressure, and air pressure. Referring to studies at Lake Mead and Lake Eucumbene, Omar and El-Bakry (1981) applied a different format of Eq. 3.10 in their estimation of evaporation from Lake Nasser, Aswan Dam. Stannard and Rosenberry (1991) credited the Bowen ratio equation to E.R. Anderson and a Lake

Fig. 3.8 Bowen ratio instrumentation site in the Everglades Nutrient Removal Project (German 2000; U.S. Geological Survey)

Hefner, Oklahoma, evaporation study. Both formats use a constant and air pressure in place of γ. The Bowen ratio estimation equation (Eq. 3.10) is presented with analysis by Reis and Dias (1998):

$$\beta = \gamma \frac{(T_s - T_a)}{(e_s - e_d)} \tag{3.10}$$

where T_s is lake surface water temperature (°C), T_a is air temperature over the lake (°C), e_s is saturation vapor pressure corresponding to T_s (kPa), and e_d is the air actual vapor pressure corresponding to T_a (kPa).

A Bowen ratio system for measuring water surface temperature, air temperature, humidity, net radiation, and heat flux is shown in Fig. 3.8.

3.6 Lidar (Light Detection and Ranging Method)

Raman lidar makes three-dimensional measurements of water vapor concentration over a surface using the Monin–Obukhov similarity theory. It samples the local time-average vertical gradient of water vapor. Local evaporation flux is calculated from this using similarity theory and supplementary measurements of friction velocity and atmospheric stability (Shuttleworth 2008).

Raman lidar application to measure water vapor concentration has been tested on a large area (0.75 km^2) at a spatial resolution of 25 m (Eichinger et al. 2000). Estimates of ET were within RMS error of 18 wm^{-2}. The application has the advantage of showing spatial variation of ET. A Los Alamos Raman lidar ET observation over corn and soybean fields showed a high degree of spatial variation of ET over a field (Eichinger et al. 2006). This technology has the potential to map spatial variation of ET on a field.

3.7 Satellite-Based Methods

The latest technology of satellite-based environmental monitoring has promising advancement for providing meteorological variable observations for large areas. Satellite-based ET estimation is presented in detail in Chaps. 10, 11, and 12.

3.8 Summary

There are many empirical models to estimate evaporation and ET. Comparison to locally measured values is the only way to gauge and improve error of estimation. In situ lysimeter installations have been a good source of ET measurements, and the need continues. Regional and water body hydrologic mass balance analysis can provide gross estimates of the ET component of the hydrology. Indirect measurements of ET are also important, and continued development of gauges is essential. Advancement in saturated and unsaturated moisture measurements will advance the science of hydrologic accounting.

Acknowledgements We would like to acknowledge Ed German from U.S. Geological Survey for taking the photographs shown in Figs. 3.6, 3.7, and 3.8.

References

Abtew W (1996) Evapotranspiration measurements and modeling for three wetland systems in South Florida. J Am Water Resour Assoc 127(3):140–147

Abtew W (2001) Evaporation estimation for Lake Okeechobee in South Florida. J Irrig Drain 127(3):140–147

Abtew W (2005) Evapotranspiration in the Everglades: comparison of Bowen ratio measurements and model estimates. In: Proceedings of the 2005 ASAE annual international meeting. ASAE, St. Joseph, MI

Abtew W, Hardee J (1993) Design of a lysimeter for a wetland environment: evapotranspiration of cattails (*Typha domingensis*). Paper presented at the 1993 international meeting of ASAE. Paper No 932516. ASAE, St. Joseph, MI

Abtew W, Obeysekera J (1995) Lysimeter study of evapotranspiration of cattails and comparison of three estimation methods. Trans ASAE 38(1):121–129

Abtew W, Anderson DL, Lindstrom LJ, Cadogan A (1998) Soil moisture and water table monitoring for irrigation and drainage decision making. In: Brown LC (ed) Drainage in the 21st century: food production and the environment. ASAE, St. Joseph, MI

Abtew W, Iricanin N, Obeysekera J (2011) Pan evaporation and potential evapotranspiration trends in South Florida. Hydrol Process 25:958–969

Crowell ML, Mtundu ND (2000) Guidelines for quality control and quality assurance of hydrologic and meteorological data. Volume 2: Data management. South Florida Water Management District, West Palm Beach

Eichinger WE, Cooper D, Kao J, Chen LC, Hipps L, Prueger J (2000) Estimation of spatially distributed latent heat flux over complex terrain from a Raman lidar. Agric Forest Meteorol 105(1–3):145–159

Eichinger WE, Cooper D, Hipps LC, Kustas WP, Neale CMU, Prueger J (2006) Spatial and temporal variation in evaporation using Raman lidar. Adv Water Resour 29(2):369–381

German ER (2000) Regional evaluation of evapotranspiration in the Everglades. Water Resources Investigations Report 00-4217. USGS, Tallahassee, FL

Gifford RM, Roderick M, Farquhar GD (2007) Evaporative demand: does it increase with global warming? Glob Change Newsl 69:21–23

Jensen ME (ed) (1974) Consumptive use of water and irrigation water requirements. Report prepared by Technical Committee on irrigation water requirements. Irrigation and Drainage Division, ASCE, New York

Omar MH, El-Bakry MM (1981) Estimation of evaporation from the lake of the Aswan High Dam (Lake Nasser) based on measurements over the lake. Agric Meteorol 23:293–308

Reis RJD, Dias NL (1998) Multi-season lake evaporation: energy-budget estimates and CRLE model assessment with limited meteorological observations. J Hydrol 208:135–147

Shuttleworth WJ (1993) Chapter 4, Evaporation. In: Maidment D (ed) Handbook of hydrology. McGraw-Hill, Inc, New York

Shuttleworth WJ (2008) Evapotranspiration measurement method. Southwest Hydrol 7(1):1–23

Stannard DI, Rosenberry DO (1991) A comparison of short-term measurements of lake evaporation using eddy correlation and energy budget methods. J Hydrol 122:15–22

Chapter 4
Energy Requirements of Dew Evaporation

Abstract Dew formation and its importance in the hydrologic cycle and energy requirements for evaporation is a scientific interest and important in high-resolution evapotranspiration modeling. In humid areas, where dew point temperature is reached at night and in the morning, significant energy is required to evaporate the dew. In this chapter, detailed methods of estimating dew and the energy required for evaporation are presented along with results from experimental study. High-resolution meteorological observations coupled with temperature and wind profile measurements were used to develop heat transfer coefficients and other parameters required to estimate sensible and latent heat. The 44 days of study of dew evaporation resulted in an estimate of 5% of energy required for daily evaporation being used to evaporate dew. The study also showed that on the average, 75 min of duration of morning net radiation is required to evaporate dew at the study site. As much as 0.5-mm daily dew evaporation was estimated.

Keywords Dew • Dewfall • Dew evaporation • Energy balance

4.1 Introduction

As air temperature falls in the evening, at times, it reaches the dew point temperature. Dew point temperature is presented in Chap. 2. As air temperature falls, it could reach a point where it cannot hold water vapor at its current level. This is the temperature at which air moisture in the form of water vapor transforms to liquid form and deposit on soil, vegetation, and other surfaces. Condensation of water vapor from the atmosphere occurs at night when total energy is negative. The condensation process results in heat energy release, a reverse of the evaporation process in the daytime. This includes both sensible and latent heat transfer. The rate of condensation is dependent on humidity, temperature, wind speed, and cloud cover. The interest in the study of dew formation and the utilization by plants dates back to the seventeenth century (Stone 1963).

W. Abtew and A. Melesse, *Evaporation and Evapotranspiration: Measurements and Estimations*, DOI 10.1007/978-94-007-4737-1_4,

The purpose of this chapter is to present the significance of energy required to evaporate dew or rainfall from vegetation surfaces. Hourly or higher-resolution evapotranspiration modeling can be improved by accounting for early morning energy requirements for drying wet leaves. Earlier work by the author is referenced, and additional material is presented. The physical approach of evapotranspiration modeling accounts for the balance and transfer of energy, momentum, and mass. In regions with high rainfall, high humidity, and frequent nights with dew formation, the amount of energy required in the morning to dry out wet surfaces requires consideration. Air saturated with vapor, on clear nights, can result in maximum dew formation at a rate of 0.07 mm h^{-1} (Monteith 1973). A precise weighing lysimeter was used to measure quantity and duration of dew and actual evapotranspiration near Deniliquin, New South Wales, Australia (Sharma 1976). Results showed that the summer months of December, January, and February had negligible dew formation. The rest of the year, dew amounted to 1.2% of class A pan evaporation, 2.5% of precipitation, and 3.9% of actual evapotranspiration. A maximum dew amount of 0.56 mm day^{-1} was recorded during the winter of 1974. Dew plays critical role in desert ecology. Measurement of dew with micro-lysimeters in the Negev Desert, Israel, resulted in dew formation rates of 0.1–0.2 mm per night (Jacobs et al. 1999). Measurements of dew on winter wheat produced deposition rates of 0.02–0.33 mm per night (Burrage 1971).

4.2 Energy Balance and Transfer Coefficients

The early morning energy balance on vegetation surfaces can be expressed by Eq. 4.1 with the assumption that advective energy is negligible (Abtew and Obeysekera 1995):

$$H + \lambda E = R_n - G \tag{4.1}$$

where H is sensible heat flux, λE is latent heat flux, R_n is net radiation, and G is heat gained or lost by the vegetation mass. In Eq. 4.1, all terms are negative for energy flow away from the leaf surface and positive for net radiation, which is an energy input. During dew formation, λE is negative indicating that condensation is opposite to evaporation.

In modeling the evaporation and condensation processes, momentum, mass, and energy transfer mechanisms have to be accounted. Shear stress, latent heat, and sensible heat fluxes are presented in general form as follows (Eqs. 4.2, 4.3, and 4.4):

$$\tau = \rho k_m \frac{du}{dz} \tag{4.2}$$

where τ is shear stress, ρ is air density, k_m is transfer coefficient for shear stress, and du/dz represents the change in wind speed with height.

$$\lambda E = \frac{\lambda \varepsilon}{P} k_{\mathrm{w}} \frac{\mathrm{d}e}{\mathrm{d}z} \tag{4.3}$$

where λ is latent heat of vaporization of water, λE is latent heat, ε is the ratio of molecular weights of water to dry air, P is atmospheric pressure, k_{w} is coefficient for latent heat transfer, and $\mathrm{d}e/\mathrm{d}z$ is vapor pressure change with height.

$$H = \rho c_{\mathrm{p}} k_{\mathrm{h}} \frac{\mathrm{d}T}{\mathrm{d}z} \tag{4.4}$$

where c_{p} is specific heat of air, k_{h} is coefficient of sensible heat transfer, and $\mathrm{d}T/\mathrm{d}z$ is temperature change with height.

The three transfer coefficients (k_{m}, k_{w}, k_{h}) are dependent on wind speed, humidity, and temperature. Surface conditions and atmospheric stability are also factors to be considered (Katul and Parlange 1992). For most applications, the three transfer coefficients are assumed to be equal (Federer 1970). The heat transfer coefficient (k_{h}) has been expressed in implicit and explicit forms. Explicit forms from various sources are presented as follows (Eqs. 4.5, 4.6, and 4.7):

$$k_{\mathrm{h}} = u_*^2 \frac{\mathrm{d}z}{\mathrm{d}u} \tag{4.5}$$

where u_* is friction velocity and $\mathrm{d}z/\mathrm{d}u$ is the inverse of the wind speed gradient (Monteith 1973).

$$k_{\mathrm{h}} = \frac{k u_* (z - d + z_{\mathrm{h}})}{\Phi_{\mathrm{h}}} \tag{4.6}$$

where k is the von Karman constant (0.41), z is height, d is displacement height, z_{h} is roughness length for heat transfer, and $_{\mathrm{h}}$ is a stability correction factor, a function of the Monin–Obukhov length (Stannard 1993).

$$k_{\mathrm{h}} = u_* \theta_* \frac{\mathrm{d}z}{\mathrm{d}T} \tag{4.7}$$

where θ_* is temperature scale and is computed by Eq. 4.8 as the inverse of the temperature gradient (Jacovides et al. 1992).

$$\theta_* = \frac{\Delta T k}{\ln\left(\frac{z_2}{z_1}\right)} \tag{4.8}$$

where ΔT is temperature difference between two heights of measurement (z_1 and z_2). Federer (1970) provided the following equation (Eq. 4.9):

$$k_{\mathrm{h}} = \frac{k u_* z}{\Phi_{\mathrm{h}}} \tag{4.9}$$

4.3 Dewfalls and Evaporation

In a study, high-resolution data was collected where relative humidity and air temperature were measured at 1- and 2-m heights in a constructed wetland in south Florida (26° 38′ N, 80° 25′ W). Wind speed and direction were measured at three heights: 1, 2.6, and 10 m (Abtew and Obeysekera 1995). A Campbell Scientific model 237 leaf wetness sensor was used to detect and measure leaf surface wetness. When the surface was wet due to dew formation or rainfall, the resistance dropped below 200 kΩ. When dry, the readings were over 200 kΩ. For illustration, 3 days (April 7, May 15, and December 24, 1994) with no recorded rainfall are presented to demonstrate dew formation, evaporation, and associated meteorological variables. The wetness in these days is attributed to dew formation. Figure 4.1 depicts leaf wetness from dewfall on three nights/mornings with no rainfall and dry out in the morning. Solar radiation, net solar radiation, and photosynthetic photon flux density (PPFD) were measured at 2-m height. Figure 4.2 depicts net solar radiation on the three mornings when dew formation and drying was observed.

From Fig. 4.2, net radiation started at 6:30 am, 7:00 am, and 8:15 am on April 7, May 15, and December 24, 1994, respectively. On the corresponding days, leaf dry out occurred at 7:30 am, 8:00 am, and 10:30 am (Fig. 4.1). December 24, 1994, had low net radiation and air temperature due to time of the year (Figs. 4.2 and 4.5). In winter months, significant amounts of solar energy would be used for evaporation from wet surfaces.

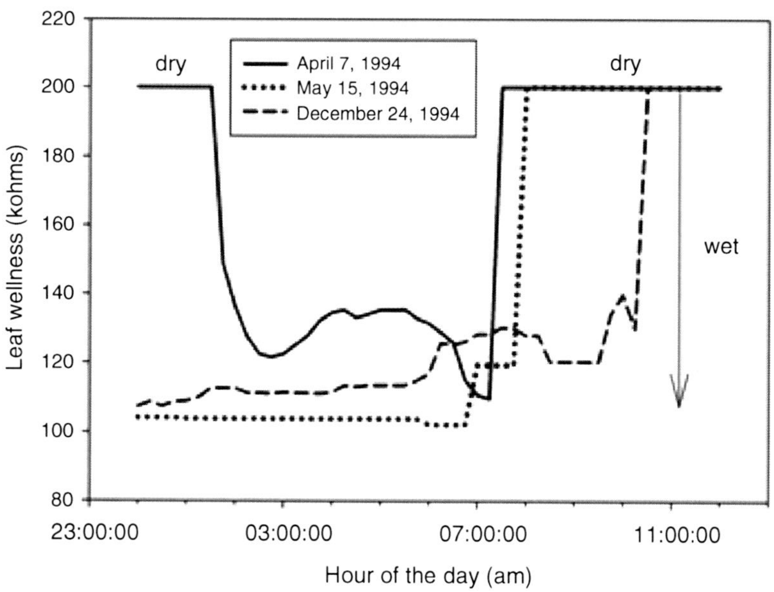

Fig. 4.1 Leaf wetness from dew and dry out from dew evaporation

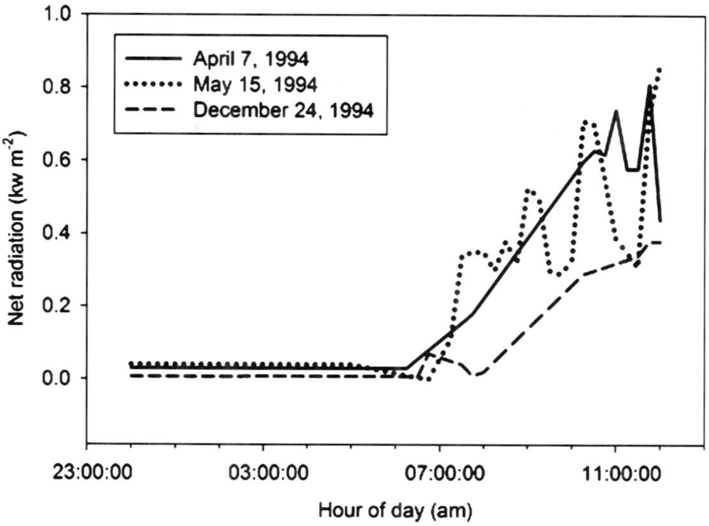

Fig. 4.2 Net radiation on three mornings

The start time for energy input for drying wet surfaces in the morning is depicted in Fig. 4.2, and the end time is shown in Fig. 4.1 when drying starts. The low energy on December 24, 1994, morning is reflected in the dew (wet leaf) delayed drying in Fig. 4.1. Relative humidity 15-min observations on the three mornings of the study also show that in general, 4:15 am to 7:15 am was 100% humidity before drying started. The delay in dew drying on December 24, 1994, is also reflected in the relative humidity observations (Fig. 4.3).

For the purpose of wind speed gradient and profile determination, wind speed, vector, and direction were measured at three heights: 1, 2.6, and 10 m. Wind parameters were sampled every 10 s and recorded as 15-min averages. All other parameters were measured at intervals of 5 min and recorded as 15-min averages. Wind speed measurements at two heights were needed for estimation of u_*, friction velocity, for computing k_h using Eqs. 4.5, 4.6, 4.7, and 4.8. Friction velocity was estimated using Eq. 4.10:

$$u_* = \frac{k(u_2 - u_1)}{\ln(z_2 - d) - \ln(z_1 - d)} \tag{4.10}$$

where k is von Karman constant, u_2 and u_1 are wind speed measurements at heights z_2 and z_1, and d is displacement height computed as a function of average vegetation height and fraction of cover (Abtew et al. 1989).

Daily mean heat transfer coefficients, k_h, computed by the four equations (Eqs. 4.5, 4.6, 4.7, and 4.8) are close to what was computed by Eq. 4.6. Therefore, Eq. 4.6 was used for the 44 days of study. Near neutral atmospheric stability was assumed in Eq. 4.7. The roughness length for heat transfer, z_h, in Eq. 4.6, was

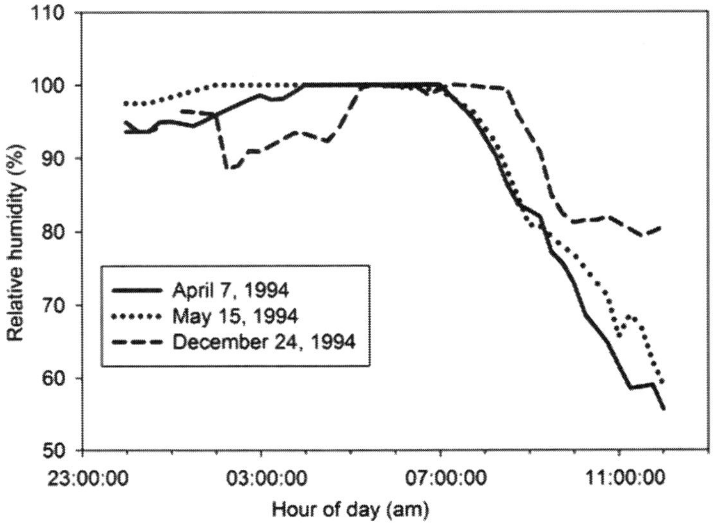

Fig. 4.3 Relative humidity for the mornings of April 7, May 15, and December 24, 1994

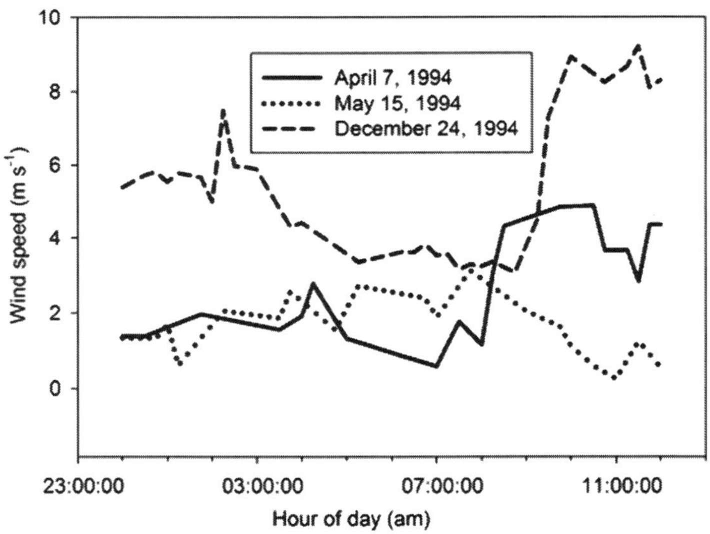

Fig. 4.4 Wind speed at 10 m for the mornings of April 7, May 15, and December 24, 1994

estimated by Eq. 4.11 (Allen et al. 1989). Figure 4.4 depicts wind speed at 10 m for the mornings of the 3 days. Wind speed increased with sunrise for the mornings of April 7 and December 24 but not on May 15, 1994:

$$z_h = 0.1 z_o \tag{4.11}$$

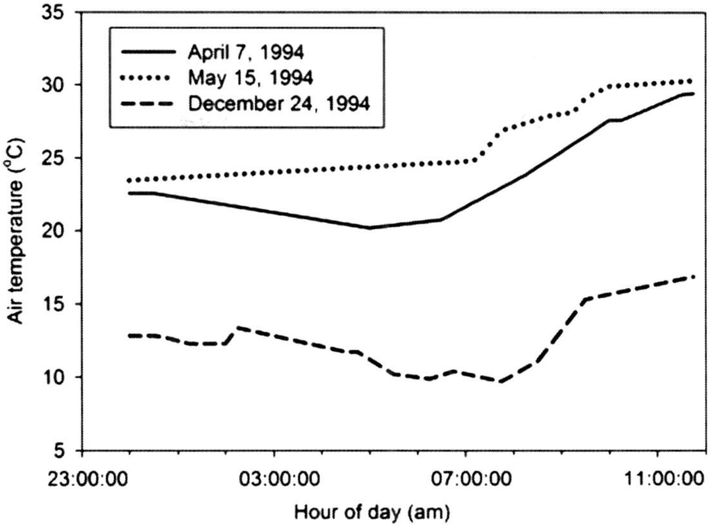

Fig. 4.5 Air temperature at 2-m height

where z_o is aerodynamic roughness (s m^{-1}) as computed in Abtew et al. (1989). Equation 4.6 is not sensitive to the roughness length for heat transfer as its relative magnitude is small. Equation 4.4 was applied to compute sensible heat (H) . Dew evaporation (E) for the duration (t_1 to t_2), the time from positive net radiation observation to leaf drying (Figs. 4.2 and 4.1), was computed by Eq. 4.12:

$$E = \frac{\int_{t_1}^{t_2} R_n \mathrm{d}t - \int_{t_1}^{t_2} H \mathrm{d}t}{\lambda} \tag{4.12}$$

where the latent of heat of vaporization of water (λ) was computed by Eq. 4.13, following Smith (1991).

$$\lambda = 2.54 - 0.002361T \tag{4.13}$$

where T is temperature at 2-m height. Figure 4.5 depicts high-resolution air temperature for the 3 days presented for illustration. The rise in air temperature in the morning corresponds to net radiation pattern and dew evaporation. Dew evaporation for the 3 days was estimated using average parameters from Table 4.1 ($\mathrm{d}T/\mathrm{d}z$, k_h). The estimates are 0.11, 0.31 , and 0.51 mm for April 7, May 15, and December 24, 1994.

From the observation, it is apparent that high-resolution evapotranspiration modeling such as hourly time interval needs to account for energy used to dry the wet leaf surface. An energy balance during the time period when wet vegetation dries out can be applied to estimate energy used to evaporate morning dew and to

Table 4.1 Measured and computed meteorology and energy balance parameters during dew evaporation, 7:00–8:15 am

Parameter	Parameter mean	Equation
$U_{10\,\mathrm{m}}$ (ms^{-1})	2.02	Observed
$U_{2.6\,\mathrm{m}}$ (ms^{-1})	1	Observed
$U_{2\,\mathrm{m}}$ (ms^{-1})	0.58	Wind profile
$U_{1\,\mathrm{m}}$ (ms^{-1})	0.02	Observed
U_* (ms^{-1})	0.1	Eq. 4.10
$T_{2\,\mathrm{m}}$ (°C)	16.6	Observed
$T_{1\,\mathrm{m}}$ (°C)	16.32	Observed
$\mathrm{RH}_{2\,\mathrm{m}}$ (%)	96.9	Observed
$\mathrm{RH}_{1\,\mathrm{m}}$ (%)	97.8	Observed
R_n (KJ m^{-2} s^{-1})	0.09	Observed
H (KJ m^{-2} s^{-1})	0.011	Eq. 4.4
λE (KJ m^{-2} s^{-1})	0.079	Eq. 4.1
E (mm)	0.15	Eq. 4.12
P (kPa)	101.6	Observed
K_h (m^2 s^{-1})	0.048	Eq. 4.6
Q_* (°C)	−0.17	Eq. 4.8

estimate depth of dewfall. Equation 4.1 expresses the early morning energy balance on vegetation mass. Assuming change in heat storage for the vegetation mass to be negligible, latent and sensible heat can be computed as net radiation is available from observations. Means of parameters from the 44 days of the study are shown in Table 4.1. The mean duration of dew evaporation was 75 min, and the mean dew evaporation from leaf surfaces was 0.15 mm. It was also assumed that wet leaf surface evaporation did not start until positive net solar radiation readings started. The time taken by dew evaporation in the morning was defined as the time interval between start of positive net radiation and leaf dry out.

At the site, a lysimeter was designed and installed inside a cattail marsh with a similar environment simulating cattail evapotranspiration under saturated conditions. The fully automated lysimeter has a surface area of 9.8 m^2 (Abtew and Hardee 1993). In the morning, the time evapotranspiration started in the lysimeter was compared to the time leaf dryness started. Several days of correspondence were observed between the time evapotranspiration started in the lysimeter and the time leaf surface dried out (Abtew and Obeysekera 1995). Generally, lysimeter evapotranspiration lagged behind leaf dryness. The reason could be that the leaf wetness sensor is fully exposed to wind and solar radiation while underlying leaves in the lysimeter take longer to dry.

4.4 Summary

Since this study employed single leaf wetness sensors, the volume of water held in thick vegetation with layers of leaves would be higher than the estimates provided in this study. Underlying leaves would take longer to dry out, resulting in more dew evaporation. In general, the nights and mornings when dew formation is favorable due to sufficient moisture in the air and dew point temperature, early morning solar radiation, will be consumed for dew evaporation. The 44 days of study of dew evaporation resulted in an estimated minimum of 5% of energy being consumed to evaporate dew. The study also showed that on the average, 75 min of morning net radiation is required to evaporate dew.

References

Abtew W, Hardee J (1993) Design of a lysimeter for a wetland environment: evapotranspiration of cattails (*Typha domingensis*). Paper No. 932516. ASAE, St. Joseph, MI

Abtew W, Obeysekera J (1995) Estimation of energy requirements of morning dew evaporation from leaf surfaces. Water Resour Bull 31(2):217–225

Abtew W, Borrelli J, Gregory JM (1989) Wind profile: estimation of displacement height and aerodynamic roughness. Trans ASAE 32(2):521–527

Allen RG, Jensen ME, Wright JL, Burman RD (1989) Operational estimates of reference evapotranspiration. Agron J 81:650–662

Burrage SW (1971) Dew on wheat. Agric Meteorol 10:3–12

Federer CA (1970) Measuring forest evapotranspiration – theory and practice. USDA Forest Service Research Paper NE-165. U.S. Department of Agriculture, Washington, DC

Jacobs AFG, Heusinkveld BG, Berkowicz SM (1999) Dew deposition and drying in a desert system: a simple simulation model. J Arid Environ 42(3):211–222

Jacovides CP, Kerkides P, Papiaoannou G, Smith FB (1992) Evaluation of the profile and the resistance method for estimation of surface fluxes and momentum, sensible and latent heat. Theor Appl Climatol 45:145–154

Katul GG, Parlange MBA (1992) Penman-Brutsaert model for wet surface evaporation. Water Resour Res 28(1):121–126

Monteith JL (1973) Principles of environmental physics. Edward Arnold, London

Sharma ML (1976) Contribution of dew in the hydrologic balance of a semi-arid grassland. Agric Meteorol 17(5):321–331

Smith M (ed) (1991) Report on the expert consultation of FAO guidelines for prediction of crop water requirements. United Nations, Rome

Stannard DI (1993) Comparison of Penman-Monteith, Shuttle-Worth-Wallace, and modified Priestley-Taylor evapotranspiration models for wildland vegetation in semiarid rangeland. Water Resour Res 29(5):1379–1392

Stone EC (1963) The ecological importance of dew. The University of Chicago Press, Chicago

Chapter 5
Vapor Pressure Calculation Methods

Abstract Evapotranspiration (ET) or water loss to the atmosphere is one of the largest components of the hydrologic cycle, and its estimation is subject to uncertainties. Most ET estimation methods depend on vapor pressure deficit estimation. Improvements in saturation vapor pressure, actual vapor pressure, and vapor pressure deficit computations contribute to reducing errors in estimating ET. Using high-resolution meteorological data, various vapor pressure computations methods were compared. High-resolution saturation vapor pressure can be computed from high-resolution meteorological data reflecting diurnal fluctuations. In the absence of high-resolution meteorological data, daily average saturation vapor pressure is best estimated from the daily 24-h average relative humidity and the 24-h average air temperature followed by the average of daily maximum and minimum air temperature. Actual vapor pressure is best estimated from the 24-h mean air temperature and relative humidity. With some error, the average of the maximum and minimum air temperature and relative humidity can be applied to estimate actual vapor pressure. In this study, application of many equations is presented with correlation of the results with "true" estimates.

Keywords Vapor pressure • Vapor pressure deficit • Vapor pressure calculation

5.1 Introduction

Most ET estimation models depend on the estimation of vapor content of the air, its capacity to hold more, and vapor pressure deficit (vpd). Vapor pressure deficit is a major factor in the rate and amount of mass transfer. The amount of water vapor in saturated air is dependent on the temperature of the mixture. The higher the temperature is, the higher the capacity to hold water vapor. Vapor pressure deficit is the difference between saturation vapor pressure and actual vapor pressure ($e_s - e_d$). Anderson (1936) and others realized early on that percent humidity by itself is not

W. Abtew and A. Melesse, *Evaporation and Evapotranspiration: Measurements and Estimations*, DOI 10.1007/978-94-007-4737-1_5,

a measure of dryness but vapor pressure deficit. The selection of equations for the computation of e_s and e_d has direct effect on the calculation of vpd for use in ET estimation models.

Jensen et al. (1990) have presented ET models that rely on vpd and discussed the commonly used vapor pressure computation methods. Evaporation estimation methods such as Penman, Penman-combination, Penman–Monteith, Van Bavel–Businger, and mass transfer models have vapor pressure components. Sadler and Evans (1989) have discussed errors of ET estimation associated in vpd computation methods. Howell and Dusek (1995) summarized the literature concerning the application of diverse methods for computing vapor content of the air. They also compared vpd computation methods for the semiarid region of the Southern High Plains (Bushland, Texas).

5.2 Comparison of Vapor Pressure Computation Methods

5.2.1 Methods

Vapor pressure (e_d) is dependent on air temperature and humidity. The capacity of air to hold moisture increases as air temperature increases and vice versa. The diurnal variation of saturation vapor pressure follows the diurnal variations of air temperature. Vapor pressure (actual) is computed from saturation vapor pressure (e_s) and relative humidity. The difference between e_s and e_d is the vapor pressure deficit (vpd), which is a driver in the rate of evaporation. Saturation vapor pressure is computed as follows (Eq. 5.1):

$$e_s = 0.611 \exp\left(\frac{17.27T}{T + 237.3}\right) \tag{5.1}$$

where e_s is saturation vapor pressure in kPa and T (°C) is 24-h average air temperature or maximum air temperature, minimum air temperature, or average of daily maximum and minimum temperature depending on the equation selected to compute actual vapor pressure (e_d).

Eight methods of e_d computations were evaluated against a "true" e_d as computed from the difference of the "true" e_s and "true" vpd. "True" e_s was computed based on Eq. 5.2 from 15-min average air temperature:

$$e_s = \frac{1}{96}\sum_{i=1}^{96} 0.611 \exp\left(\frac{17.27T_i}{T_i + 237.3}\right) \tag{5.2}$$

where T_i is average air temperature in °C for the 15-min time interval, i, for the day. The "true" vpd was computed from "true" e_s and average relative humidity (RH). Daily vpd was computed, as shown in Eq. 5.3 (Monteith 1973).

$$\text{vpd} = e_s\left(1 - \frac{\text{RH}}{100}\right) \quad (5.3)$$

A "true" e_d is the difference between "true" e_s and "true" vpd. Six commonly used e_d estimation methods are presented as follows, and the daily estimates are compared to the "true" estimate. An equation used for estimating saturation vapor pressure is applied to estimate actual vapor pressure by using the daily minimum air temperature (Eqs. 5.4, 5.5, 5.6, 5.7, 5.8, and 5.9):

$$e_d = 0.611 \exp \frac{17.27 T_{\min}}{(T + 237.3)} \quad (5.4)$$

where e_d is actual vapor pressure in kPa and $T_{\min}$ is the day's minimum temperature in °C.

$$e_d = e_s(T_{\text{avg24}})\frac{\text{RH}_{\text{avg24}}}{100} \quad (5.5)$$

where $e_s(T_{\text{avg24}})$ is saturation vapor pressure computed from the daily average air temperature (°C) and RH_{avg24} is the daily average humidity in percent.

$$e_d = e_s(T_{\min})\frac{\text{RH}_{\max}}{100} \quad (5.6)$$

where $e_s(T_{\min})$ is saturation vapor pressure computed from the daily minimum air temperature (°C) and $\text{RH}_{\max}$ is the daily maximum humidity in percent.

$$e_d = e_s(T_{\max})\frac{\text{RH}_{\min}}{100} \quad (5.7)$$

where $e_s(T_{\max})$ is saturation vapor pressure computed from the daily maximum air temperature (°C) and $\text{RH}_{\min}$ is the daily minimum humidity in percent.

$$e_d = e_s(T_{\text{avg2}})\frac{\text{RH}_{\text{avg2}}}{100} \quad (5.8)$$

where $e_s(T_{\text{avg2}})$ is saturation vapor pressure computed from the average of the daily minimum and maximum air temperature (°C) and RH_{avg2} is the daily average humidity in percent computed as average of the daily minimum and maximum relative humidity.

$$e_d = \frac{1}{2}\frac{e_s(T_{\min})}{100}\text{RH}_{\max} + \frac{1}{2}\frac{e_s(T_{\max})}{100}\text{RH}_{\min} \quad (5.9)$$

where e_d is computed as average of two methods presented earlier.

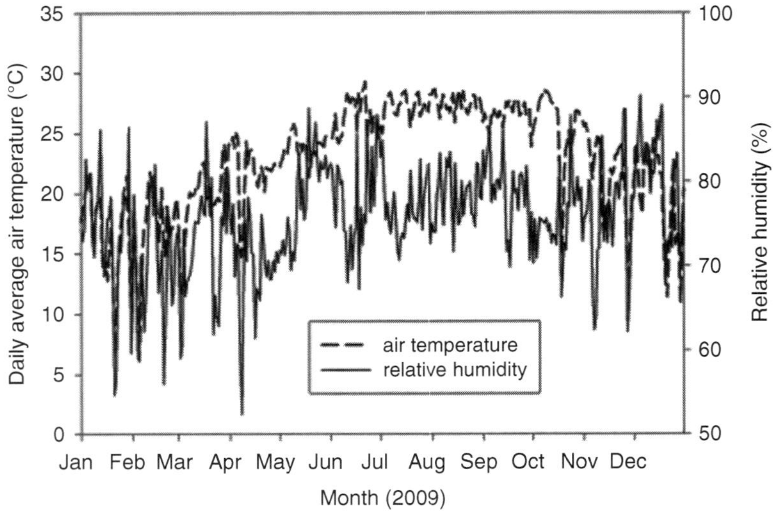

Fig. 5.1 High-resolution air temperature and relative humidity observations at a site in south Florida

5.2.2 *Results*

Mean daily saturation vapor pressure, actual vapor pressure, and vapor pressure deficit are dependent on air temperature, humidity data, and the selected equation. High-resolution air temperature and humidity data can be used to estimate "true" vapor pressure and compare the results to estimates from different equations and inputs. Figure 5.1 depicts air temperature and humidity daily variations at a site in south Florida (26° 38″ N, 80° 25″ W, elevation 3 m NGVD29) used to compute "true" vapor pressure. Data was acquired at a 2 m height with a HMP35C probe sampling every 5 min and recording 15-min average. The average air temperature and humidity for year 2009 were 22.9 °C and 76%, respectively.

Mean daily saturation vapor pressure, actual vapor pressure, and vapor pressure deficit estimates vary with the method of calculation. Methods of mean daily vpd computation that are more influenced by daytime temperature and humidity conditions overestimate mean daily vpd but better estimate mean daytime vpd. Stockle and Kiniry (1990) have reported that plant radiation-use efficiency is related to vpd. The daytime vpd computations could be important in plant water use, radiation-use studies, and plant growth models. Methods of vpd estimation are presented in Cuenca and Nicholson (1982), Sadler and Evans (1989), Jensen et al. (1990), and Howell and Dusek (1995).

Daily mean "true" e_s, e_d, and vpd as computed from 15-min time interval air temperature and humidity data are shown in Fig. 5.2 for a sample year. Estimation of daily average vpd depends on the estimation of the saturation vapor pressure and the actual vapor pressure. Both parameters depend on the selection of computation

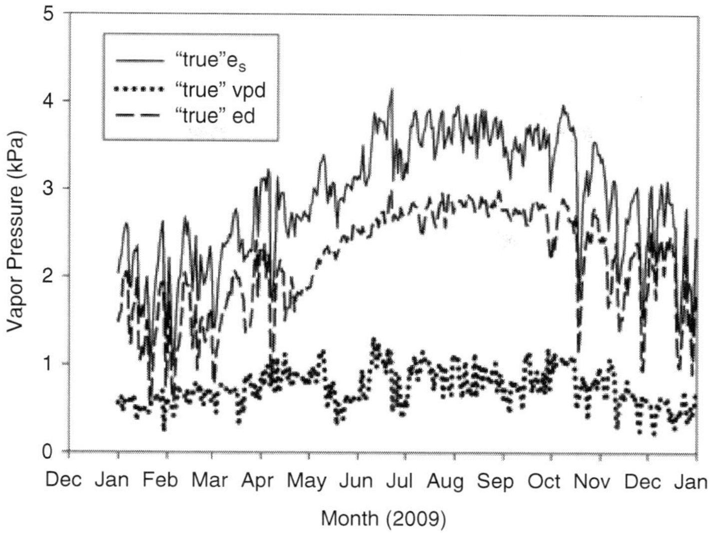

Fig. 5.2 "True" saturation (e_s), actual (e_d) vapor pressure, and vpd daily distribution for a site in south Florida

Table 5.1 Comparison of mean daily saturation vapor "true" e_s with values estimated by three equations

e_s	Mean (kPa)	Std (kPa)	a	b	r	$S_{y/x}$(kPa)
X "True" e_s	2.98	0.71	–	–	–	–
Y e_s (T_{avg24})	2.87	0.71	−0.03	1	1	0.04
Ye_s (T_{avg2})	2.94	0.76	−0.15	1.06	0.98	0.11
$Y1/2(e_s(T_{max}) + e_s(T_{min}))$	3.02	0.76	−0.06	1.06	0.98	0.12

equation. Table 5.1 depicts comparison of the "true" e_s computed from Eq. 5.2 with estimates of e_s computed with Eq. 5.1 with 24-h average temperature (T_{avg24}), with average of maximum and minimum daily temperature (T_{avg2}), and e_s as average of e_s computed using minimum daily temperature (T_{min}) and maximum daily temperature (T_{max}). The table compares means, standard deviation, and standard errors of estimates of the different methods and the "true" values. A regression statistic is provided to measure how well the different methods estimate vapor pressure.

The "true" average e_s for the analysis year of 2009 was 2.98 kPa with standard deviation of 0.71 kPa. A previous study reported a mean of 2.94 kPa and standard deviation of 0.63 kPa from 808 days of analysis (February 1993–April 1995) from the same site (Abtew 1995). The same study reported a mean daytime-to-nighttime vpd ratio of 8.7. As a comparison, a daytime-to-nighttime vpd ratio for the low-humidity, higher latitude and altitude region of the Southern High Plains (Bushland, Texas) was 3.21 ($n = 706$), as derived from Howell and Dusek (1995).

From Table 5.1, it is shown the method that uses the 24-h daily mean air temperature provides the best estimate compared to the other methods followed by

Table 5.2 Comparison of mean daily vapor pressure (actual) "true" e_d with estimates from five methods

e_d	Mean (kPa)	Std (kPa)	a	b	r	$S_{y/x}$ (kPa)
X "True" e_d	2.17	0.59	–	–	–	–
Y e_d (Eq. 5.4)	2.29	0.67	−0.06	1.09	0.96	0.13
Y e_d (Eq. 5.5)	2.19	0.60	0.01	1	1	0.03
Y e_d (Eq. 5.6)	2.08	0.63	−0.16	1.03	0.98	0.09
Y e_d (Eq. 5.7)	2	0.58	−0.03	0.94	0.96	0.16
Y e_d (Eq. 5.8)	2.19	0.6	0.01	1	1	0.03
Y e_d (Eq. 5.9)	2.04	0.59	−0.09	0.98	0.99	0.13

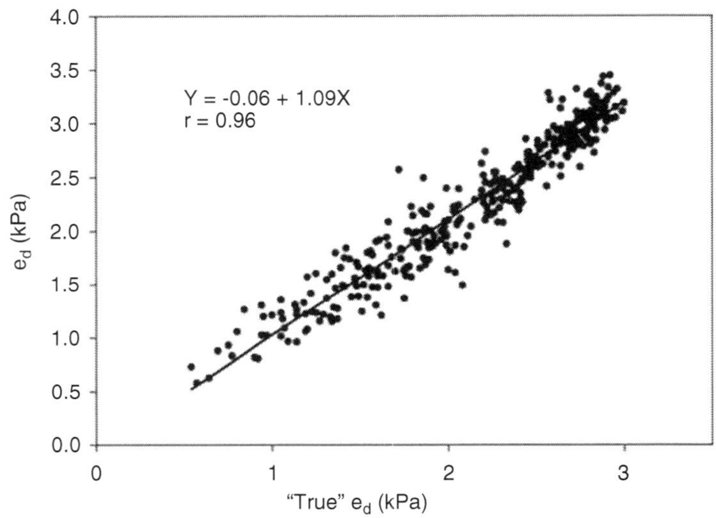

Fig. 5.3 Comparison of e_d computed with Eq. 5.4 and the "true" e_d

the method that uses average of the daily minimum and maximum air temperature. The least preferred method is average e_s computed from daily minimum and maximum air temperatures.

The "true" e_d was computed as a difference of the "true" e_s and the "true" vpd as computed with average daily humidity in Eq. 5.3. The mean and standard deviations were 2.19 and 0.6 kPa, respectively. Table 5.2 depicts comparison of the "true" e_d with e_d computed by Eqs. 5.4, 5.5, 5.6, 5.7, 5.8, and 5.9.

Comparison of actual vapor pressure computed with Eq. 5.4 and the "true" actual daily average vapor pressure is shown in Fig. 5.3 with a correlation coefficient of 0.92. The mean and standard deviation from this method are 2.29 and 0.67 kPa. The standard error of estimation is 0.13 kPa.

Comparison of actual vapor pressure computed with Eq. 5.5 and the "true" actual daily average vapor pressure is shown in Fig. 5.4 with a correlation coefficient of close to 1. The mean and standard deviation from this method are 2.19 and 0.60 kPa. The standard error of estimation is 0.03 kPa.

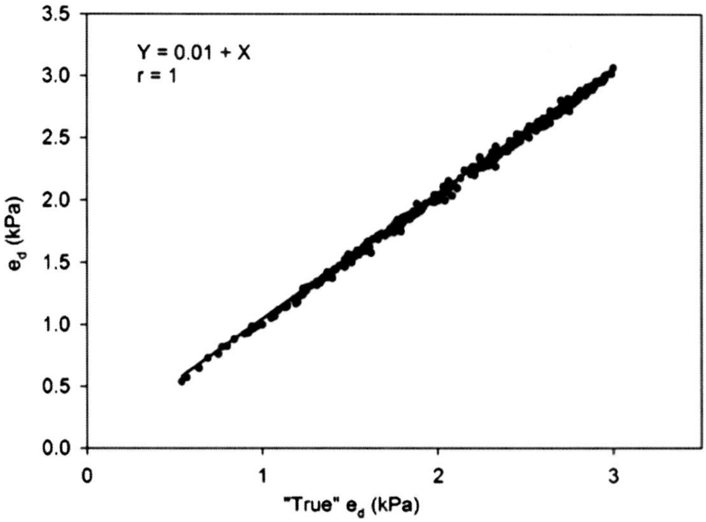

Fig. 5.4 Comparison of e_d computed with Eq. 5.5 and the "true" e_d

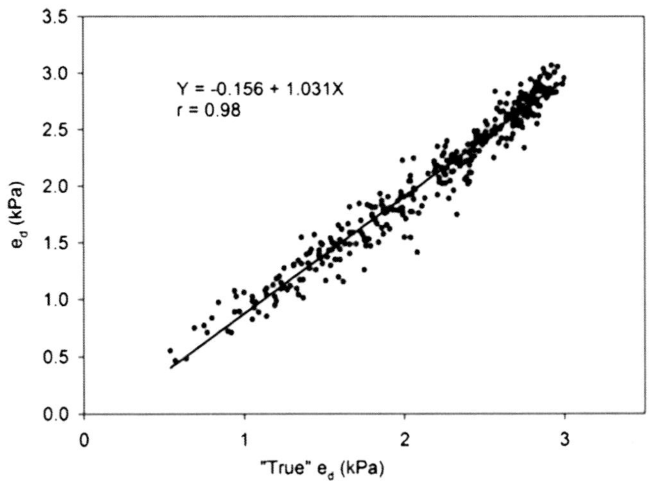

Fig. 5.5 Comparison of e_d computed with Eq. 5.6 and the "true" e_d

Comparison of actual vapor pressure computed with Eq. 5.6 and the "true" actual daily average vapor pressure is shown in Fig. 5.5 with a correlation coefficient of 0.96. The mean and standard deviation from this method are 2.08 and 0.63 kPa. The standard error of estimation is 0.09 kPa.

Comparison of actual vapor pressure computed with Eq. 5.7 and the "true" actual daily average vapor pressure is shown in Fig. 5.6 with a correlation coefficient of 0.92. The mean and standard deviation from this method are 2.00 and 0.58 kPa. The standard error of estimation is 0.16 kPa.

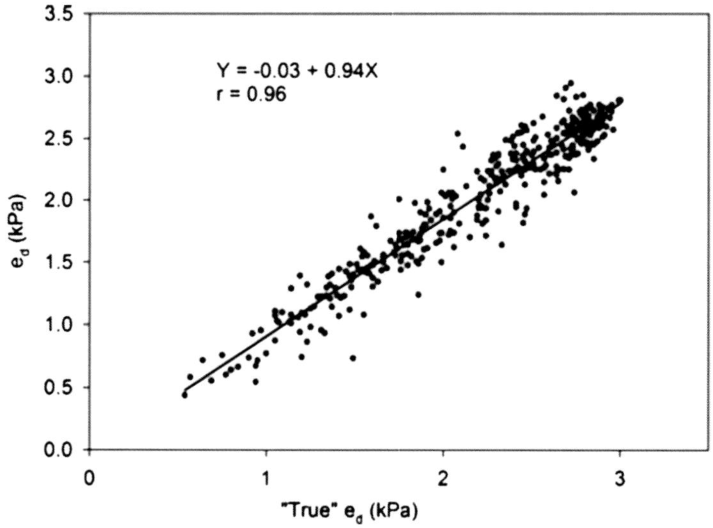

Fig. 5.6 Comparison of e_d computed with Eq. 5.7 and the "true" e_d

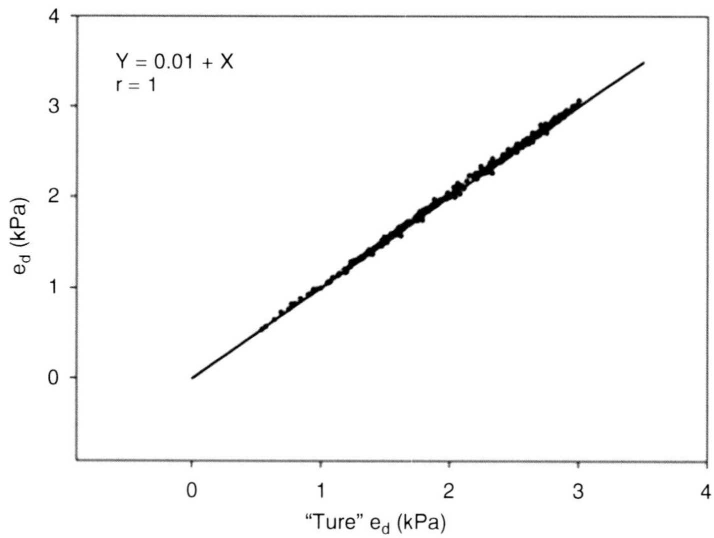

Fig. 5.7 Comparison of e_d computed with Eq. 5.8 and the "true"

Figure 5.7 depicts comparison of e_d computed with Eq. 5.8 and the "true" $e_{d.}$ The mean and standard deviation from this method are 2.19 and 0.6 kPa. The standard error of estimation is 0.03 kPa.

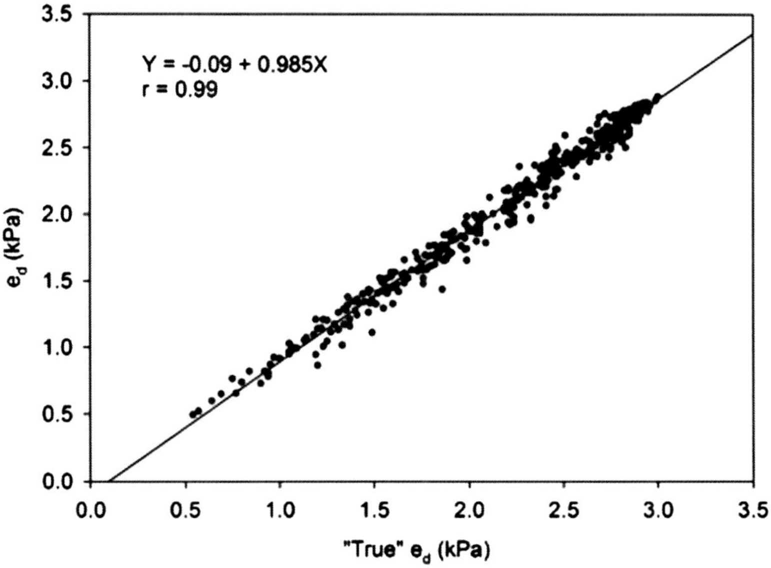

Fig. 5.8 Comparison of e_d computed with Eq. 5.9 and the "true" e_d

Figure 5.8 depicts comparison of e_d computed with Eq. 5.9 and the "true" e_d. The mean and standard deviation from this method are 2.04 and 0.59 kPa. The standard error of estimation is 0.13 kPa.

The best estimate of actual vapor pressure, e_d, is from Eq. 5.8 where both temperature and humidity are averages of the daily maximum and minimum respective readings and Eq. 5.5 where 24-h average temperature and humidity are needed.

5.3 Summary

Vapor pressure deficit is a parameter required in ET estimation equations. Understanding and evaluation of the relative accuracy of saturation and actual vapor pressure computation equations are essential for best result in ET estimation. In most cases, the high-resolution meteorological data used to compute the "true" vapor pressure deficit may not be available. A previous analysis based on 808 days and the current analysis for the humid and warm region of south Florida provided similar results. High-resolution saturation vapor pressure can be computed from high-resolution meteorological data reflecting diurnal fluctuations. In the absence of high-resolution meteorological data, daily average saturation vapor pressure is best estimated from the daily 24-h average temperature or the average of daily maximum and minimum air temperature.

Actual vapor pressure is best estimated from the 24-h mean air temperature and relative humidity. With some error, the average of the maximum and minimum air temperature and relative humidity can be applied to estimate actual vapor pressure when only such data is available.

References

Abtew W (1995) Evaluation of vapor pressure calculation methods – South Florida. DOR 225. South Florida Water Management District, West Palm Beach, FL

Anderson DB (1936) Relative humidity or vapour pressure deficit. Ecology 17(2):277–282

Cuenca RH, Nicholson MT (1982) Application of Penman equation wind function. J Eng Drain 108(1):13–23

Howell TA, Dusek DA (1995) Comparison of vapour-pressure-deficit calculation methods – Southern High Plains. J Irrig Drain 121(2):191–198

Jensen ME, Burman RD, Allen RG (ed) (1990) Evapotranspiration and irrigation water requirements. Manuals and reports on engineering practice No. 70. ASCE, New York

Monteith JL (1973) Principles of environmental physics. Edward Arnold, London

Sadler EJ, Evans DE (1989) Vapour pressure deficit calculations and their effect on the combination equation. Agric Forest Meteorol 49:55–80

Stockle CO, Kiniry JR (1990) Variability in crop radiation-use-efficiency associated with vapour-pressure-deficit. Field Crop Res 25:171–181

Chapter 6
Evaporation and Evapotranspiration Estimation Methods

Abstract Estimation in spite of measurement is the common approach to acquire ET data for most applications. Selection of a method for a specific application requires evaluation of methods with respect to the accuracy needed, available input data, and cost of data generation. Methods vary by complexity and input data requirement. In cases where simple methods provide reasonable estimates, adaptation of such methods could be a cost-effective way of acquiring ET data. In this chapter, several open water evaporation and ET estimation methods are provided with application to a region. Methods are organized from the simplest to the most complex with evaluation of input data requirements. Measured, derived, and estimated input parameters for the application of the Penman–Monteith method are presented in detail with experimental measurements of resistance terms.

Keywords Evaporation • Evapotranspiration • Lake evaporation estimation methods • Evapotranspiration estimation methods • Penman–Monteith

6.1 Introduction

Most ET estimation models are empirical. Usually, the models are statistical correlations of evaporation with minimum, maximum, or mean meteorological parameters. Performances differ from location to location, and sometimes application to a specific location requires recalibration. The Penman–Monteith method is a complex method that is closest to a physical model, accounting for mass, momentum, and energy transfer with external and internal resistance and conductance terms incorporated. ET estimation method selection depends on the availability and quality of meteorological data and site features. The subject of meteorological data quality is discussed in detail in Chap. 2. Simple methods require fewer input parameters and could satisfy needs in many regions where intensive data collection networks are not available and are costly.

W. Abtew and A. Melesse, *Evaporation and Evapotranspiration: Measurements and Estimations*, DOI 10.1007/978-94-007-4737-1_6,

6.2 Simple Methods

6.2.1 *Pan Method*

Estimating lake evaporation from pan evaporation is the simplest method but has many challenges. Evaporation from a small metallic pan usually placed on dry site is higher than evaporation from a lake. Advective energy, heat storage difference, and higher vapor pressure deficit due to the site environment result in higher evaporation. A coefficient, K_p, is used to reduce pan evaporation to estimate lake evaporation (Eq. 6.1). Reference crop evapotranspiration (Eq. 6.2) is also estimated from pan evaporation using a coefficient (C_{et}). Coefficients K_p and C_{et} are dependent on the type of pan, environment at the site, and pan operation. Wide ranges of these coefficients are available indicating that pan evaporation measurements are affected by site-specific factors. These factors include site location, type of pan, quality of measurements, and operations and maintenance. Comparison of pan evaporation data from seven sites around Lake Okeechobee in south Florida resulted in pan coefficients ranging from 0.64 to 0.95, demonstrating that each pan is influenced by local environment and operations (Abtew 2001; Abtew et al. 2011). Spatial variation of the pan evaporation to surface water evaporation ratio over the United States for May through October (warm months) is mapped with a range of 0.64–0.88 (Farnsworth et al. 1982). On this map, pan coefficients for south Florida range from 0.72 to 0.74. Reference crop coefficient, C_{et}, is dependent on meteorological conditions such as wind speed and humidity, and the values range from 0.35 to 0.85 (Jensen 1974):

$$E_o = K_p E_{pan} \tag{6.1}$$

$$E_{tp} = C_{et} E_{pan} \tag{6.2}$$

Historical pan evaporation data are usually plagued with outliers, gaps, and data of questionable quality. Variations in pan evaporation data within relatively small distances indicate the challenges of acquiring consistent observations from pans. The wide range of pan coefficients tends to overemphasize the shortcoming of pan data (Shuttleworth 1993; Abtew et al. 2011).

Pan evaporation data from south Florida were analyzed to see if the data quality is sufficient to determine evaporation trends (Abtew et al. 2011). A total of nine pan evaporation sites with varying lengths of record from 1916 to 2009 were used for this analysis. Missing data less than a week were estimated mainly by interpolation. Months and years with too many missing days were excluded. In many cases where several daily data were available as a cumulative value on the last day of record, the values were redistributed equally for each day of accumulation. The maximum annual record at a site was 210 cm, and the minimum record was 119 cm. The mean pan annual evaporation for all sites was 156 cm with a standard deviation of 18.5 cm. The distance between gauges was a maximum of 109 km and a minimum

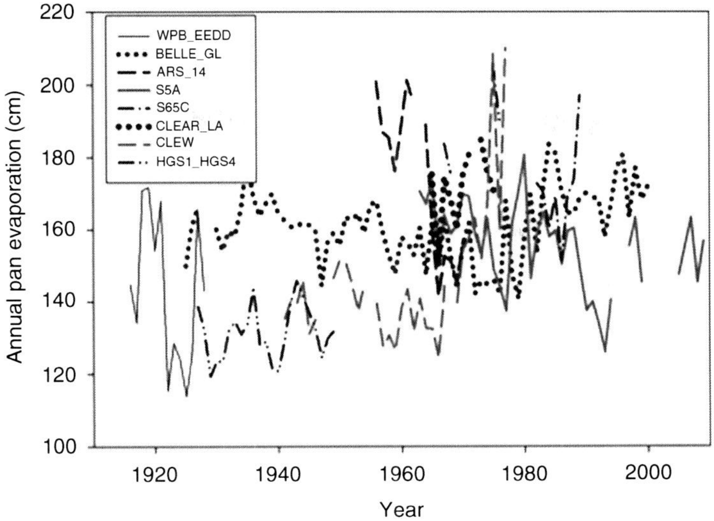

Fig. 6.1 Annual pan evaporation data from eight sites in south Florida showing variation in measurements and recording

of 0.2 km. These ranges of records reflect the challenges of acquiring good quality pan evaporation observations rather than actual variation in the parameter in a subregion. Provided water is available for evaporation and the site environment and operation are similar, the annual spatial variation in evaporation should not be as large as recorded at these sites. Probably, these challenges are common at pan evaporation sites at other parts of the world. Even if there was no error in observations, the role of the microclimate and environment differences between sites and differences in site operations could produce varying results within a subregion (Abtew et al. 2011). Figure 6.1 depicts monthly pan evaporation from eight sites in south Florida showing variation in measurement and recording. Screening of data for quality and assembling of data from many sites could provide a set of pan evaporation data for applications. Lake evaporation and crop evapotranspiration estimation from pan evaporation are discussed in Chaps. 3 and 8. Well-installed, maintained, and operated evaporation pans could consistently provide good-quality data, and evaporation can be estimated with locally calibrated pan coefficients.

6.2.2 *Temperature-Based Methods*

6.2.2.1 Blaney–Criddle Method

Temperature-based ET estimation methods are the simplest methods. The most commonly applied temperature-based evapotranspiration estimation method is the

Table 6.1 Mean daily percentage (p) of annual sunshine hours for different latitudes

North lat.	Jan	Feb	Mar	Apr	May	Jun	Jul	Aug	Sep	Oct	Nov	Dec
South lat.	Jul	Aug	Sep	Oct	Nov	Dec	Jan	Feb	Mar	Apr	May	Jun
50	0.19	0.23	0.26	0.31	0.34	0.36	0.36	0.32	0.28	0.24	0.20	0.18
48	0.20	0.23	0.26	0.31	0.34	0.36	0.35	0.32	0.28	0.24	0.21	0.19
46	0.20	0.23	0.27	0.30	0.33	0.35	0.34	0.32	0.28	0.24	0.21	0.19
44	0.21	0.24	0.27	0.30	0.33	0.35	0.34	0.31	0.28	0.25	0.22	0.20
42	0.21	0.24	0.27	0.30	0.33	0.34	0.33	0.31	0.28	0.25	0.22	0.20
40	0.22	0.24	0.27	0.30	0.32	0.34	0.33	0.31	0.28	0.25	0.22	0.21
–	–	–	–	–	–	–	–	–	–	–	–	–
35	0.23	0.25	0.27	0.29	0.31	0.33	0.32	0.30	0.28	0.25	0.23	0.22
30	0.23	0.25	0.27	0.29	0.31	0.32	0.31	0.30	0.28	0.26	0.24	0.23
25	0.24	0.26	0.27	0.29	0.30	0.31	0.30	0.29	0.28	0.26	0.25	0.24
20	0.25	0.26	0.27	0.28	0.30	0.30	0.30	0.29	0.28	0.26	0.25	0.25
15	0.26	0.26	0.27	0.28	0.29	0.29	0.29	0.29	0.28	0.27	0.26	0.25
10	0.26	0.27	0.27	0.28	0.29	0.29	0.29	0.28	0.27	0.27	0.26	0.26
5	0.27	0.27	0.27	0.28	0.28	0.28	0.28	0.28	0.27	0.27	0.27	0.27
0	0.27	0.27	0.27	0.27	0.27	0.27	0.27	0.27	0.27	0.27	0.27	0.27

Extracted from Doorenbos and Pruitt (1977)

Blaney–Criddle method. The measured climatic variable input is air temperature, and the equation is as follows (Eq. 6.3):

$$\mathrm{ET_o} = p(0.46\,T_{\mathrm{avg}} + 8) \tag{6.3}$$

where $\mathrm{ET_o}$ is reference crop evapotranspiration (mm/day) average for the month, T_{avg} is mean daily temperature (°C) for the month, and p is mean daily percentage of annual daytime hours for the month (Table 6.1).

The FAO (Food and Agricultural Organization, UN) Blaney–Criddle method is a modified Blaney–Criddle that accounts for the effect of other weather parameters on crop water requirements (Doorenbos and Pruitt 1977). The method accounts for temperature, relative humidity, sunshine hours, site elevation, and wind speed to estimate reference crop evapotranspiration. The FAO Blaney–Criddle method has been in use in the western states of Nevada, Washington, Idaho, Oregon, and California to estimate irrigation requirements on a state-wide basis (Allen and Pruitt 1986). The equation is given as follows (Eq. 6.4):

$$\mathrm{ET_o} = c(p(0.46T_{\mathrm{avg}} + 8.13))\left\{1 + 0.1\frac{\mathrm{elev}}{1{,}000}\right\} \tag{6.4}$$

where $\mathrm{ET_o}$ is estimated evapotranspiration in mm day^{-1} from grass reference crop (8–15 cm tall and well watered) for the month of consideration, T_{avg} is mean daily temperature in °C for the month, p is mean daily percentage of total annual daytime hours for the month and latitude, and c is adjustment factor which depends on minimum relative humidity, wind speed, and sunshine hours. To avoid the use of graphs and tables to interpolate the correlation factor, c, two coefficients, a and b (Eqs. 6.5 and 6.6), were formulated to replace c (Frevert et al. 1983).

$$a = 0.0043\mathrm{RH}_{\min} - \frac{n}{N} - 1.41 \tag{6.5}$$

where $\mathrm{RH}_{\min}$ is average minimum relative humidity for the month and n/N is mean actual to possible sunshine ratio.

$$b = 0.819 - 0.00409\mathrm{RH}_{\min} + 1.07\frac{n}{N} + 0.0656u_{\mathrm{day}}\frac{0.00597\mathrm{RH}_{\min}}{N}n - 0.000597\mathrm{RH}_{\min}u_{\mathrm{day}} \tag{6.6}$$

where u_{day} is daytime wind speed in m s^{-1} for the month. In cases where sunshine data is not available but solar radiation data is available, Eq. 6.7 can be used to deduct n/N (Jensen 1974).

$$\frac{n}{N} = 2.08\frac{R_{\mathrm{s}}}{R_{\mathrm{A}}} - 0.48 \tag{6.7}$$

where R_{s} is solar radiation and R_{A} is extraterrestrial radiation. A demonstration of the application of the modified Blaney–Criddle method (Eq. 6.4) to estimate reference evapotranspiration in the Everglades Agricultural Area in south Florida is shown in Table 6.2.

The annual reference evapotranspiration estimate is 1,459 mm. The seasonal pattern follows the seasonal evapotranspiration pattern of the region, but this reference evapotranspiration is 5.6% higher than the potential evapotranspiration, as computed by the simple Abtew method. The Blaney–Criddle reference evapotranspiration estimates and the simple Abtew method potential evapotranspiration compare closely from December to April. The Blaney–Criddle estimates for the remaining months are higher (Fig. 6.2).

Other temperature-based ET estimation methods include the Thornthwaite method where monthly potential ET is estimated from mean monthly air temperature, daytime hours, and 12-month sum of heat index (Jensen 1974).

6.2.2.2 Hargreaves–Samani Method

The Hargreaves–Samani method is not truly a temperature-based method because it has a radiation term in it. Since measurement is not needed for the extraterrestrial radiation (R_{A}), this method may be classified as a temperature-based method. The Hargreaves equation is given by Eq. 6.8 (Hargreaves and Samani 1985):

$$\mathrm{ET_r} = aR_{\mathrm{A}}(T_{\max} - T_{\min})^{0.5}(T_{\mathrm{avg}} + 17.8) \tag{6.8}$$

Table 6.2 Application of the modified Blaney–Criddle method to estimate reference evapotranspiration in south Florida

Month	R_s (kW m^{-2})	R_A (kW m^{-2})	n/N	RH$_{min}$ %	T_{avg} (°C)	p (%)	U@ 10 M (m s^{-1})	Day/night u ratio	u_{day} (m s^{-1})	a	b	ET$_o$ (mm day^{-1})
Jan	0.14	0.27	0.63	58	17.5	0.24	3.89	1.65	4.84	−1.79	1.19	2.86
Feb	0.17	0.32	0.63	56	18.6	0.26	3.96	1.89	4.93	−1.79	1.21	3.35
Mar	0.21	0.38	0.68	54	20.4	0.27	4.09	1.89	5.09	−1.86	1.28	4.20
Apr	0.24	0.43	0.66	54	22.3	0.29	3	2	3.74	−1.84	1.22	4.59
May	0.25	0.46	0.64	54	24.7	0.30	2.5	1.63	3.11	−1.82	1.18	5.11
Jun	0.23	0.47	0.52	60	26.3	0.31	2.38	2.06	2.96	−1.67	1.03	4.77
Jul	0.23	0.46	0.53	60	26.9	0.30	2.81	2.11	3.50	−1.69	1.06	4.93
Aug	0.21	0.44	0.51	61	27.2	0.29	2.19	2.06	2.73	−1.66	1.01	4.46
Sep	0.19	0.39	0.50	63	26.6	0.28	2.09	1.98	2.60	−1.64	0.98	3.91
Oct	0.17	0.33	0.61	61	24.8	0.26	2.26	1.95	2.81	−1.75	1.08	3.77
Nov	0.15	0.28	0.66	60	21.1	0.25	3.28	1.79	4.08	−1.81	1.16	3.29
Dec	0.13	0.25	0.63	64	19.2	0.24	3.11	1.67	3.87	−1.76	1.09	2.68

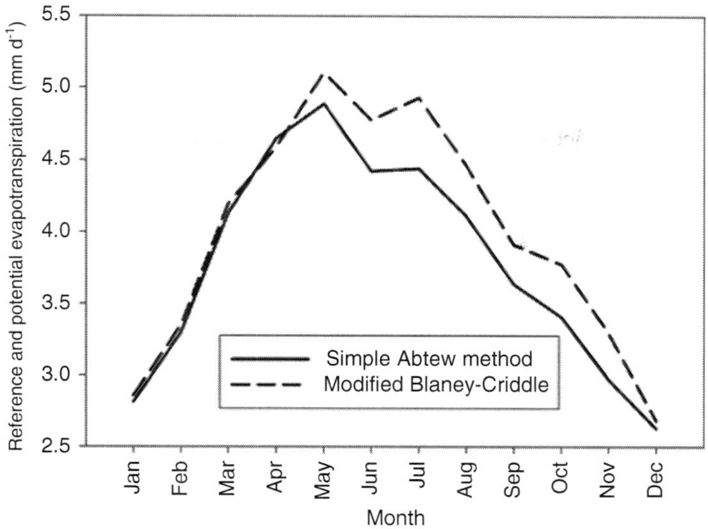

Fig. 6.2 Modified Blaney–Criddle method reference ET and simple Abtew method potential evapotranspiration for south Florida

where ET_r is grass reference ET (mm day^{-1}), T_{max} and T_{min} (°C) are maximum and minimum daily air temperature, T_{avg} is mean daily air temperature (average of daily maximum and minimum), and R_A is extraterrestrial radiation (mm day^{-1}). This method has been applied or tested in many places and widely published.

6.2.3 Radiation-Based Methods

In parts of the world where solar radiation explains most of the variation in evaporation and evapotranspiration, a simple equation may be calibrated to estimate ET from one variable, solar radiation.

6.2.3.1 The Simple Abtew Method

The simple Abtew method (Eq. 6.9) has been applied to estimate lake evaporation, wetland evapotranspiration, and potential evapotranspiration. This equation was developed from open water evaporation and wetland evapotranspiration lysimeter studies in south Florida:

$$\mathrm{ET} = K_1 \frac{R_s}{\lambda} \tag{6.9}$$

where ET is daily wetland evapotranspiration or shallow open water evaporation or potential evapotranspiration (mm day^{-1}), R_s is solar radiation (MJ m^{-2} day^{-1}), λ is latent heat of vaporization of water (MJ kg^{-1}), and K_1 is a dimensionless coefficient (0.53). The mm day^{-1} unit is derived from the fact that a kilogram of water is 1,000 cc (10^6 mm^3) and a square meter is 10^6 mm^2.

Application of this method is shown in Chaps. 7 and 8. The simple Abtew method has been successfully applied in many parts of the world. Evaporation from Lake Ziway in the Ethiopian Rift Valley was estimated with this method, and results were comparable to the energy balance and Penman equation (Melesse et al. 2009). Satisfactory results of reference ET estimation for the Fogera flood plain in Ethiopia, with the simple method, are reported with adjustment of the K_1 to 0.48 (Enku et al. 2011). The simple Abtew method was applied to estimate ET in Gansu province, northwest China, with recalibrated coefficients (Zhai et al. 2009). The simple method was applied to estimate evaporation from Lake Titicaca, South America. It was found to be the best method compared to eight evaporation models (Delclaux and Coudrain 2005). Comparative application of the simple method further demonstrates its usefulness. In an effort to identify the most relevant approach to calculate potential evapotranspiration for use in daily rainfall–runoff models, 27 potential ET models were compared for stream flow simulation from 308 catchments in France, the United States, and Australia. Each potential ET model estimate was applied to four continuous daily lumped rainfall–runoff models, and the simple Abtew method had a comparable goodness-of-fit measure (Oudin et al. 2005).

Shoemaker and Sumner (2006) applied the simple Abtew method to estimate potential evapotranspiration from open water, saw grass, and bullrush marsh and compared it to the Priestley–Taylor and Penman methods. Out of the eight sites of measurement, the simple method had the smallest standard error for two sites. The low cost of monitoring needed for this method was pointed out as a positive attribute compared to other methods that require more parameters. Xu and Singh (2000) evaluated various radiation-based methods for calculating evaporation and concluded that the simple Abtew method, referenced as the simple Abtew equation, can be used when available data is limited to radiation data. The simple method is applicable to remote sensing where the input, solar radiation, is acquired through satellite observations (Jacobs et al. 2002).

In a U.S. Geological Survey (USGS) study, nine sites in the Everglades of south Florida were instrumented with sensors to determine evapotranspiration from different features using the Bowen ratio–energy balance method (German 2000). Figure 6.3 shows the nine USGS sites and site characteristics where evapotranspiration was measured with the Bowen ratio–energy balance method. Field data is available with varying lengths of record, from 1996 to 2000. The field instrumentation had net radiometer, pyranometer, wind speed and direction sensors, air temperature and humidity sensors, rain gauge, storage battery, solar panel, data logger, and cellular phone. Pictures of a site with instrumentation are shown in Chap. 7. The Bowen ratio–energy balance method is a micrometeorological method for measurement of evaporation (latent heat) with an approximate accuracy of 10%

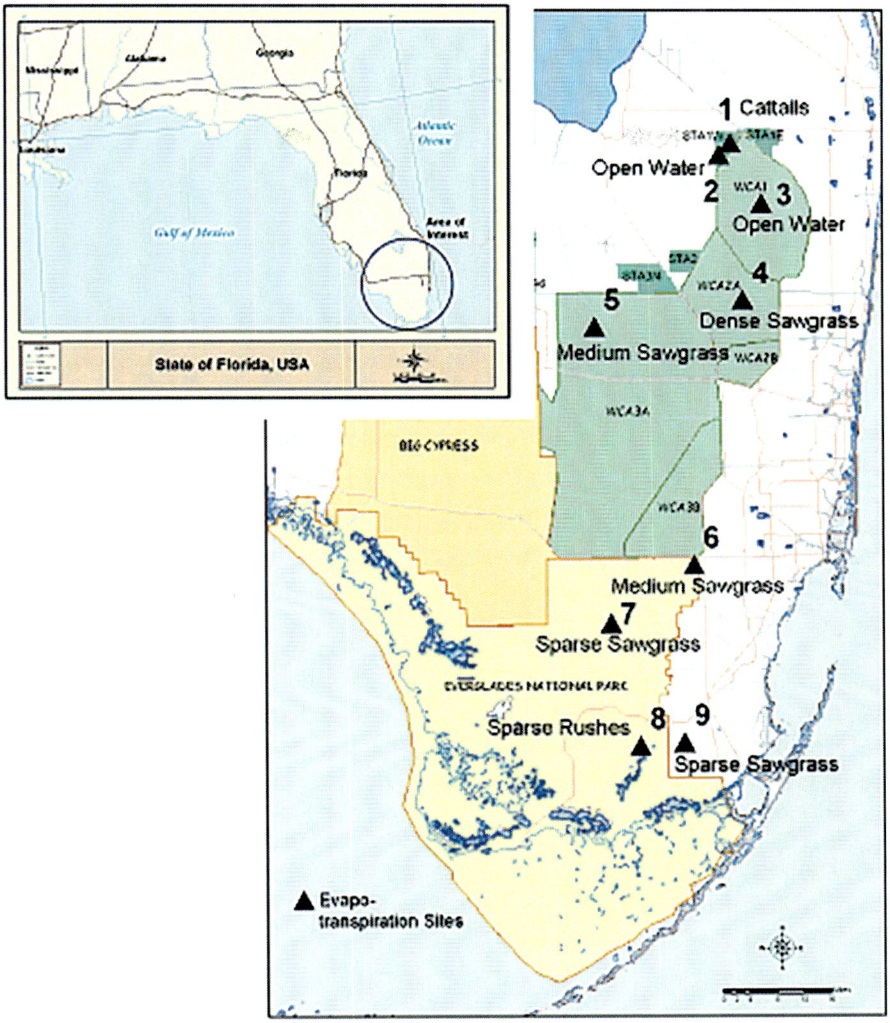

Fig. 6.3 Wetland evapotranspiration study sites in south Florida (German 2000; U.S. Geological Survey; Abtew 2005)

(Dugas et al. 1991). Details of the Bowen ratio ET measurement are given in Chaps. 3 and 7. Mean Bowen ratio ET measurement and estimates by the simple Abtew method are shown in Fig. 6.4 for seven of the sites. The mean square error for all sites is 0.06 mm, showing a very good estimation.

Lake Ziway is located in the Ethiopian Rift Valley with an average surface area of 490 km^2 at an elevation of 1,636 m msl. Monthly and annual average Lake Ziway evaporation estimates have been published. The estimates vary from method to method of evaporation estimation. Annual lake evaporation estimates by Coulomb

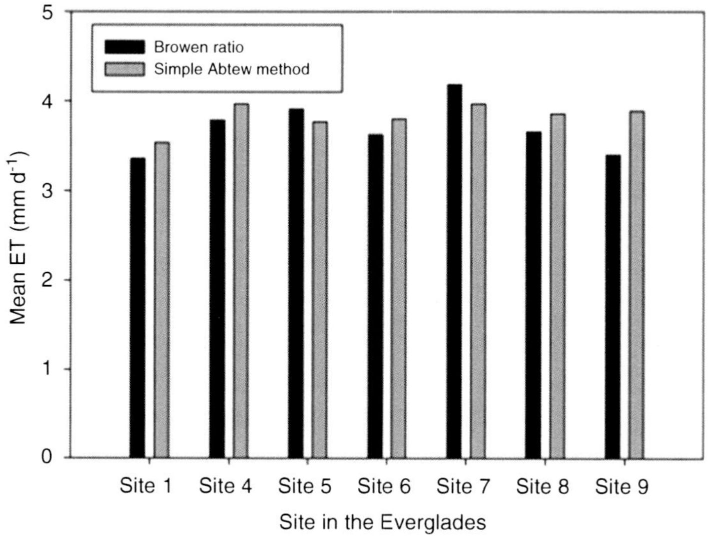

Fig. 6.4 Comparison of Bowen ratio wetland ET measurements and Simple Abtew method estimates at seven sites in the Everglades

et al. (2001) estimated with the energy balance, Penman, and Complementary Relationship Lake Evaporation (CRLE) methods were 1,777, 1,875, and 1,728 mm, respectively. The coefficient for the simple method can be adjusted to the results of the three methods or to the one that is believed to be closer to the true estimates. As an illustration, the energy balance and the simple Abtew method were compared with the K value in Eq. 6.9 revised to 0.57 from 0.53. The results are shown in Fig. 6.5. Detail of the energy balance method application for lake evaporation is presented in Chap. 8.

6.2.3.2 Makkink Method

The Makkink method (1957) is classified as radiation-based method, although it requires mean air temperature (°C), relative humidity, and air pressure (mb) to calculate the slope of saturation vapor pressure curve (Δ) and the psychrometric constant (γ). The original Makkink equation is given as follows (Eq. 6.10):

$$\mathrm{ET} = 0.61 \frac{\Delta R_{\mathrm{s}}}{(\Delta + \gamma)\lambda} \tag{6.10}$$

where ET is potential evapotranspiration from grass (cm day^{-1}), R_{S} is solar radiation in cal cm^{-2} day^{-1}, Δ is the slope of saturation vapor pressure curve (mb °C), γ is the psychrometric constant (mb °C), and λ is latent heat of vaporization (cal gm^{-1}). Δ, γ, and λ are computed by Eqs. 6.11, 6.12, and 6.13 (Maidment 1993).

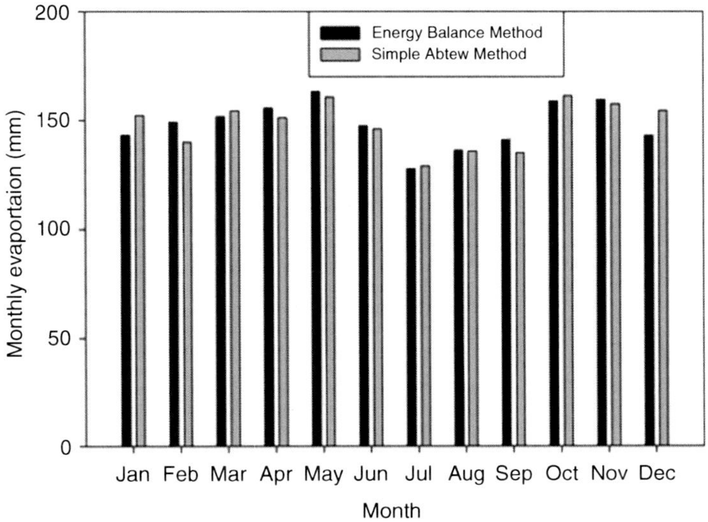

Fig. 6.5 Comparison of energy balance and simple Abtew method evaporation estimates for Lake Ziway, Ethiopia

$$\Delta = \frac{4098 e_{\mathrm{s}}}{(237.3 + T)^2} \tag{6.11}$$

where e_{s} (kPa) is saturation vapor pressure and T is given air temperature (°C).

$$\gamma = 0.0016286 \frac{P}{\lambda} \tag{6.12}$$

where P (kPa) is atmospheric pressure.

$$\lambda = 2.501 - (0.00236 * T_{\mathrm{s}}) \tag{6.13}$$

where T_{s} (°C) is surface temperature of water.

For south Florida, the Makkink method to estimate potential evapotranspiration was calibrated to the simple Abtew method and is shown in Eq. 6.14. ET for 2007 was 1,330 and 1,322 mm for the Makkink and simple Abtew methods, respectively. Comparison of daily estimates by the two methods is depicted in Fig. 6.6:

$$\mathrm{ET} = 0.743 \frac{\Delta R_{\mathrm{s}}}{(\Delta + \gamma)\lambda} \tag{6.14}$$

where ET is in mm day^{-1}, Δ and γ are in kPa °C^{-1}, R_{s} is in MJ m^{-2} day^{-1}, and λ is MJ kg^{-1}. Equation 6.14 is close to what was proposed by Hansen (1984) in the Netherlands with a coefficient of 0.7.

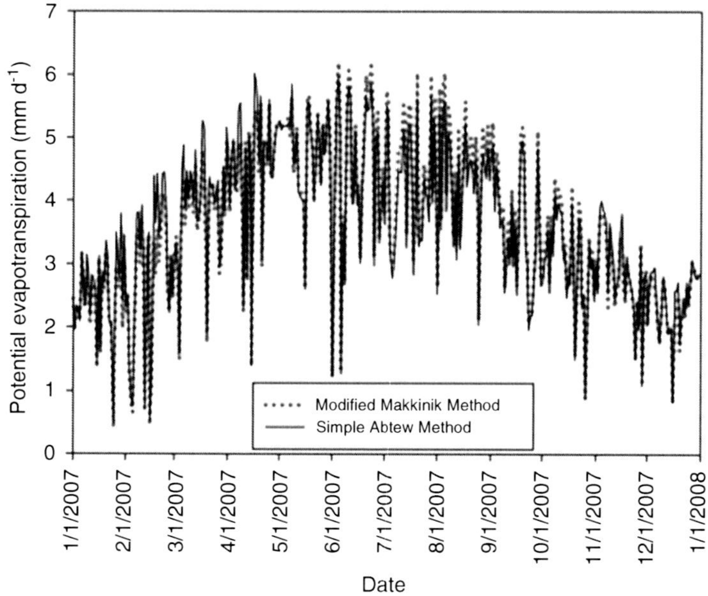

Fig. 6.6 Modified Makkink and simple Abtew methods potential evapotranspiration estimation for south Florida

6.2.3.3 Priestley–Taylor Method

The Priestley–Taylor method is similar to the Makkink method (1957), but net solar radiation (R_n) is used instead of solar radiation (R_s). Since R_n is smaller than R_s, the coefficient in Priestley–Taylor is higher ($\alpha = 1.26$) for compensation. The Priestley–Taylor method is expressed by Eq. 6.15. Measuring R_n is more problematic than measuring R_s, as discussed in Chap. 2:

$$\mathrm{ET} = 1.26\frac{\Delta R_n}{(\Delta + \gamma)\lambda} \tag{6.15}$$

where other terms are similar to Eq. 6.14.

ET estimates with the Priestley–Taylor method reflect R_n input data quality, and the application is limited by this data availability. Figure 6.7 depicts Makkink, simple Abtew, and Priestley–Taylor methods application for 1 year for south Florida. The coefficient alpha (α) was 1.18 based on previous work where the method was calibrated to lysimeter measurements (Abtew and Obeysekera 1995). From Fig. 6.7, the seasonal pattern of the Priestley–Taylor estimates is different than expected. ET rate increased in late summer and fall while it is expected to decline. The cause could be R_n data quality. Application of this method is further provided in Chaps. 7 and 8.

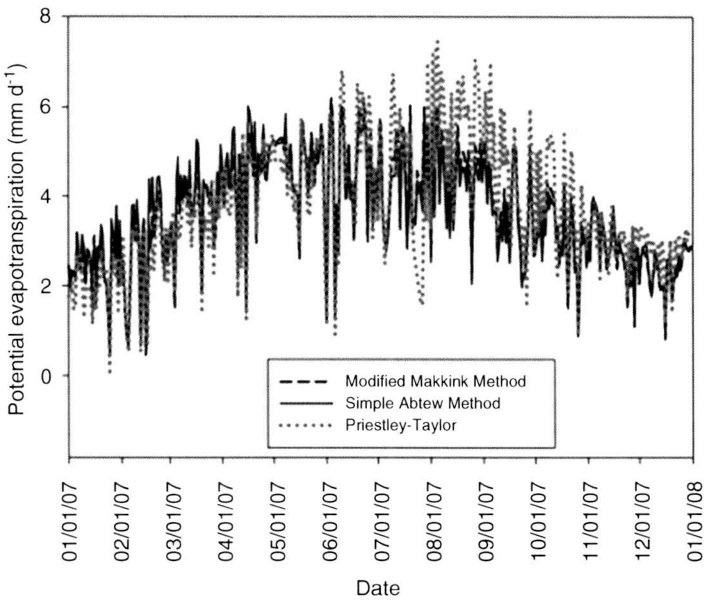

Fig. 6.7 Modified Makkink, simple Abtew, and Priestley–Taylor methods potential evapotranspiration estimation for south Florida

6.2.3.4 Turc Method

Methods that use both solar radiation and air temperature attempt to explain more ET variation by adding a second input. One of these methods is the Turc method adjusted for different units than the original equation (Eq. 6.16):

$$\mathrm{ET_p} = K_2 \frac{(23.89 R_s + 50) T_{avg}}{(T_{avg} + 15)} \tag{6.16}$$

where $\mathrm{ET_p}$ is potential evapotranspiration in mm day^{-1}, K_2 is coefficient (0.013), R_s is solar radiation in MJ m^{-2} day^{-1}, and T_{avg} is average air temperature (°C). The original Turc method estimates are lower in the first half of the year in south Florida (Fig. 6.8). In a previous study (Abtew 1996), using daily maximum air temperature (T_{max}) instead of average temperature provided a better fit to measured data in south Florida (Fig. 6.8).

Estimates for 2007 were 1,322, 1,291, and 1,390 mm for simple Abtew, Turc, and modified Turc methods. Application of the modified Turc method is presented in Chaps. 7 and 8.

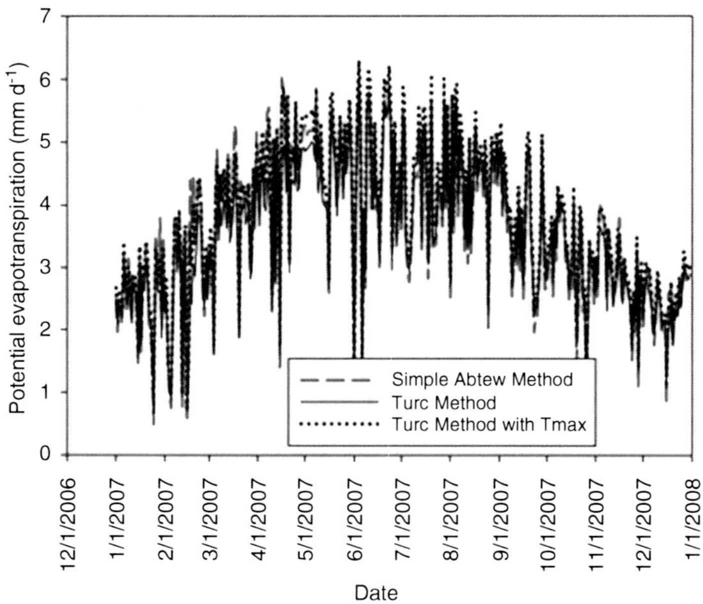

Fig. 6.8 Simple Abtew, Turc, and modified Turc methods for potential evapotranspiration estimation for south Florida

6.2.4 Solar Radiation–Maximum Temperature Method

The solar radiation–maximum temperature method was developed by the author based on lysimeter studies in south Florida reflecting radiation and maximum air temperature to explain a large portion of variation in ET in south Florida and similar environments (Abtew 1996). The method is presented by Eq. 6.17 where K_3 is a coefficient with a dimension (°C):

$$\mathrm{ET} = \frac{1}{K_3}\frac{R_s T_{\max}}{\lambda} \tag{6.17}$$

Figure 6.9 depicts a comparison of daily potential evapotranspiration (evaporation) estimates by the simple Abtew and solar radiation–maximum temperature methods. With a K_3 value of 53.5°C, ET for 2007 for the two methods was 1,322 and 1,323 mm for the year, respectively. Further application of this method is given in Chaps. 7 and 8.

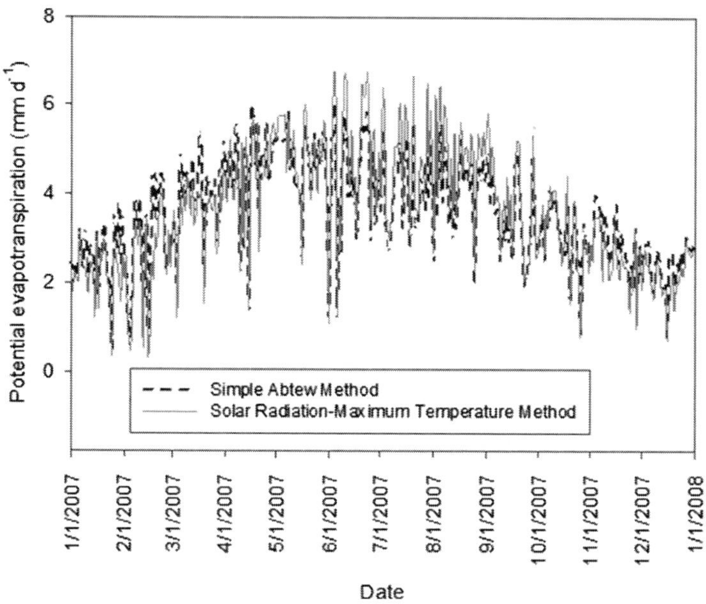

Fig. 6.9 Simple Abtew and solar radiation–maximum temperature methods potential evapotranspiration estimation for south Florida

6.2.5 *Mass Transfer Method*

The mass transfer method is based on the vapor pressure gradient from the water surface to the air above and vapor transport by wind. The common form of the formulation is given in Eq. 6.18:

$$E = k_{\mathrm{m}}\, u e_{\mathrm{s}} - e_{\mathrm{a}} \tag{6.18}$$

where E is evaporation from water surface, k_{m} is a mass transfer limiting term, u is wind speed, e_{s} is saturation vapor pressure at water surface, and e_{a} is actual vapor pressure of the air.

Details of application of the mass transfer method are presented in Chap. 8. In Chap. 8, application of the mass transfer method to a lake with lake surface water temperature and temperature of the air above the water is presented. The estimates are compared to energy balance evaporation estimation for a lake. Application was for 1 month. It is shown that when the vapor pressure deficit and wind speed are high, the mass transfer method gives high estimates beyond the energy available to sustain such an evaporation rate. At lower vapor pressure deficit, the method provides evaporation estimates that are too low. The method does not account for available energy. That is why the Penman method better described the evaporation process by combining mass transfer and energy balance components of the evaporation process.

6.3 Complex Methods

6.3.1 Energy Balance Methods

Details and application of the energy balance method are presented in Chaps. 4, 7, 8, and 10. The energy balance method accounts for energy inflow ($\text{Energy}_{\text{in}}$), energy outflow ($\text{Energy}_{\text{out}}$), and change in energy storage ($\Delta\text{Energy}_{\text{s}}$) but does not include available moisture and mechanism of mass transfer. *e* is errors. A simplified form of the energy balance method is shown by Eq. 6.19:

$$\text{Energy}_{\text{in}} - \text{Energy}_{\text{out}} = \Delta\text{Energy}_{\text{s}} \pm e \qquad (6.19)$$

The vertical energy balance at the water surface of a lake is expressed by Eq. 6.20 dropping the advective energy term:

$$\lambda E = R_{\text{n}} - H - G \qquad (6.20)$$

where λE is latent heat flux, H is sensible heat flux, and G is heat gained or lost. λ is latent heat of vaporization of water and E is evaporation. G is computed from temperature change and heat storage terms, R_{n} is measured, and H is estimated by equations that involve temperature and wind speed gradients. Details of dew evaporation, wetland evapotranspiration, and lake evaporation with the energy balance method are presented in Chaps. 4, 7, and 8.

6.3.2 The Penman Method

The Penman method is the basis for most preferred methods of evapotranspiration estimation at this time. Howard Penman in 1948 developed an equation to describe evaporation from an open water surface. The Penman equation was complete in describing the evaporation process because it has a moisture availability component, mass transfer component, and required energy for evaporation component. It requires daily mean temperature, wind speed, relative humidity, and solar radiation. Penman's equation incorporates concepts from other equations. Dalton's equation of mass moisture flux is a function of vapor pressure deficit, wind speed, and surface resistance. The resistance offered for water molecules to leave the water surface and move into the air is a function of air density, specific heat of air, psychrometric constant, latent heat of vaporization, surface resistance, and wind speed. With respect to the energy needed for evaporation, net solar radiation is divided between sensible heat and latent heat (evaporation), assuming no energy loss or gain to the ground. Sensible heat loss or gain results in change of temperature.

6.3.2.1 Mass Transfer (Sink Strength)

As shown by Dalton's equation, earlier attempts to formulate evaporation focused on mass transfer and aerodynamic components, as shown in Eq. 6.21 (Penman 1948):

$$E = (e_{ss} - e_{dd})f(u) \tag{6.21}$$

where E is evaporation per unit time, e_{ss} is vapor pressure at the evaporating surface, e_{dd} is vapor pressure in the atmosphere above, and $f(u)$ is a function of the horizontal wind. Depending on the units used for vapor pressure and wind speed, various equations had been calibrated with coefficients to estimate evaporation from vapor pressure gradient and wind speed. Application of the mass transfer equation to lake evaporation is given in Chap. 8. This approach lacks accounting for energy required for evaporation and subject to influence by wind speed and vapor pressure gradient only.

Although there could be a vapor pressure gradient, the presence of resistance at the water and air interface was realized early on. Momentum, mass, and energy transfer from a surface to the air above are a complex phenomenon. Air flowing over a surface develops a logarithmic profile as a result of a drag created by the surface (Monteith 1973; Abtew et al. 1989). Although vapor pressure deficit could exist between the surface and the air above, there are forces that resist vapor molecules from leaving the surface. On a smooth surface, the logarithmic wind velocity profile breaks close to the surface as a result of interaction with surface roughness, and a small layer of laminar flow develops transitioning to turbulent flow above. The reaction to the surface resistance force is shear stress force over the surface created from wind speed gradient. Appreciation of the complex nature of the tiny layer and forces involved is presented in detail by Monteith (1973).

Figure 6.10 illustrates the unmodified and modified turbulent layers over a rough surface with a distinct fraction of cover and boundary layer over the roughness objects. The density or fraction of cover (F_c) and height of the roughness objects (h) determines the thickness of the boundary layer and changes in the wind profile (Abtew et al. 1989). The momentum flux is highest in the unmodified flow layer, followed by the modified flow layer, and least close to the roughness surface. Eddies or still air may exist in between the roughness objects below height $d + z_o$. On a smooth flat surface, the laminar layer should be small on top of the surface. Roughness objects can be rigid or nonrigid (crop). Undulations as waves on open water act as roughness and affect the wind profile. Roughness height for water waves can be estimated (Abtew 2001). Details on wind profile are presented in Chap. 2.

Attempts to expand the aerodynamic influence in evaporation resulted in equations with more coefficients. A simplified form is shown in Eq. 6.22 referred by Penman as sink strength (Penman 1948):

$$E = 0.376(e_s - e_d)u_2^{0.76} \tag{6.22}$$

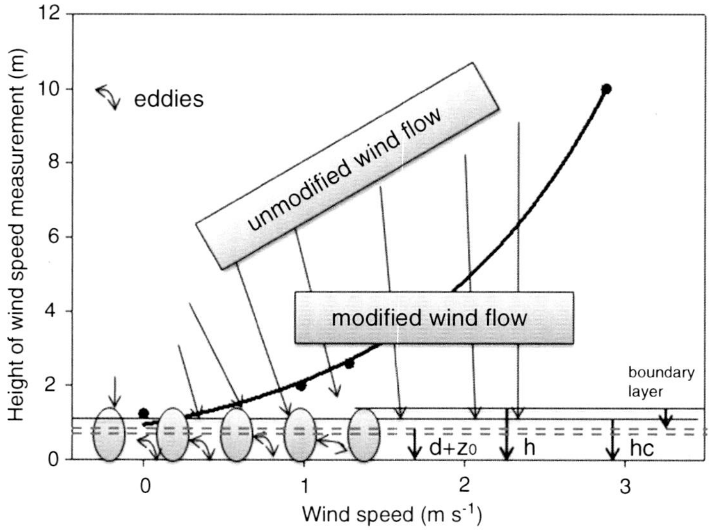

Fig. 6.10 Boundary layers when wind flows over a rough surface (roughness height (h_c), displacement height (d), and aerodynamic roughness (z_o))

where E is evaporation in mm day^{-1}, e_s and e_d are in mm mercury, and u_2 is wind speed at 2-m height measured in mph. Equation 6.22 was applied for a month period for estimating evaporation from Lake Okeechobee in south Florida. Water surface temperature was used to compute e_s (saturation vapor pressure), and air temperature was used to compute e_d (actual vapor pressure). This method is sensitive to changes in vapor pressure deficit.

Figure 6.11a depicts a comparison of evaporation computed by Eq. 22 (Penman sink strength) and the simple Abtew method. Figure 6.11b depicts daily average wind speed and vapor pressure deficit used in the calculation. Equation 6.22 was applied with the same coefficients, and the total evaporation for the month was 133 mm while the simple Abtew method gave 145 mm.

6.3.2.2 Combination of Sink Strength and Energy Balance

Penman combined the sink strength and energy balance methods to develop the Penman evaporation equation (Eq. 6.23):

$$E = \frac{\Delta R_n + \gamma \delta e f(u)}{(\Delta + \gamma)} \tag{6.23}$$

where E is latent heat of flux of evaporation (kW m^{-2}), Δ is slope of the vapor pressure curve (kPa °C^{-1}), R_n is net radiation (kW m^{-2}), δe is vapor pressure deficit

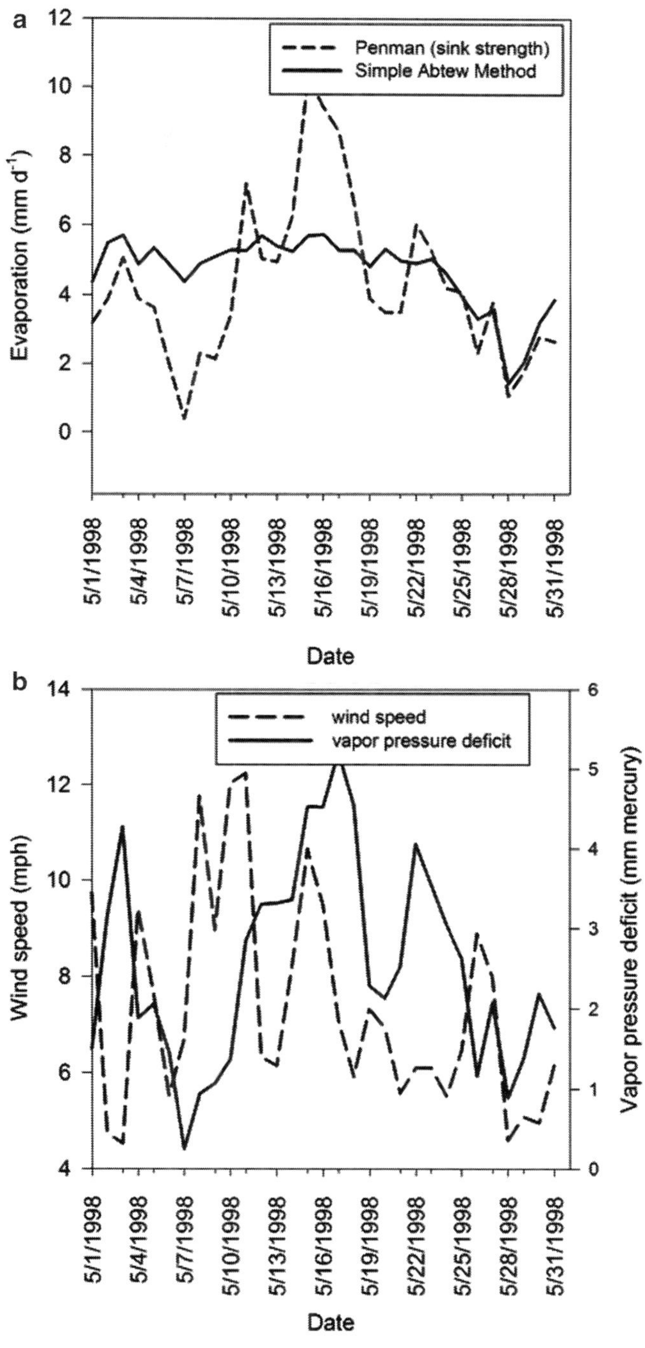

Fig. 6.11 (**a**) Penman (sink strength) and simple Abtew method evaporation estimation in south Florida; (**b**) wind speed and vapor pressure deficit

(kPa), $f(u)$ is wind function (m s^{-1}), and γ is psychrometric constant (kPa °C^{-1}). The wind function, $f(u)$, is expressed by Eq. 6.24 (Allen et al. 1989):

$$f(u) = 6.43(a_w + b_w u_2) \tag{6.24}$$

where u_2 is wind speed at 2-m height (m s^{-1}) and a_w and b_w are coefficients computed on daily basis for south Florida by Eqs. 6.25 and 6.26 (Abtew and Obeysekera 1995; Abtew 1996).

$$a_w = 0.10 + 0.2\exp\left\{-\left[\frac{J - 173}{58}\right]^2\right\} \tag{6.25}$$

$$b_w = 0.04 + 0.2\exp\left\{-\left[\frac{J - 243}{80}\right]^2\right\} \tag{6.26}$$

where J is day of the year. Application of the Penman combination method in south Florida is shown in Fig. 6.12a where estimates are compared to the simple Abtew method estimates. The simple Abtew method does not have mass transfer or sink strength component. The Penman combination method annual ET estimate (1,374 mm) is greater by 3.8% compared to the simple Abtew method (1,322 mm). Generally, during the dry season (May through November), the simple Abtew method estimates are higher. During the wet, humid months, the Penman method has higher estimates. Figure 6.12b depicts solar radiation (R_s) used in the simple method and net solar radiation (R_n) used in the Penman method.

Figure 6.12c depicts daily vapor pressure deficit and wind speed at 2-m height. It is clearly shown that the Penman method is sensitive to vapor pressure deficit.

6.3.2.3 The Penman–Monteith Method

The Penman–Monteith (P–M) equation is the closest to a physical evapotranspiration estimation model and applicable at shorter time periods than a day. Energy balance, momentum transfer, and mass transfer are accounted, and internal and external resistance or conductance to the evapotranspiration process is accounted. In this section, the P–M equation is presented with details of each parameter or coefficient used. The P–M equation has been accepted as the standard to compute reference evapotranspiration (Eq. 6.27):

$$\mathrm{ET} = \frac{1}{\lambda}\frac{\Delta(R_n - G) + \rho c_p(e_a - e_d)\frac{1}{r_a}}{\Delta + \gamma\left(1 + \frac{r_c}{r_a}\right)} \tag{6.27}$$

where ET is evapotranspiration in mm day^{-1}, Δ is the slope of the vapor pressure curve (kPa °C^{-1}), γ is psychrometric constant (kPa °C^{-1}), R_n is net radiation

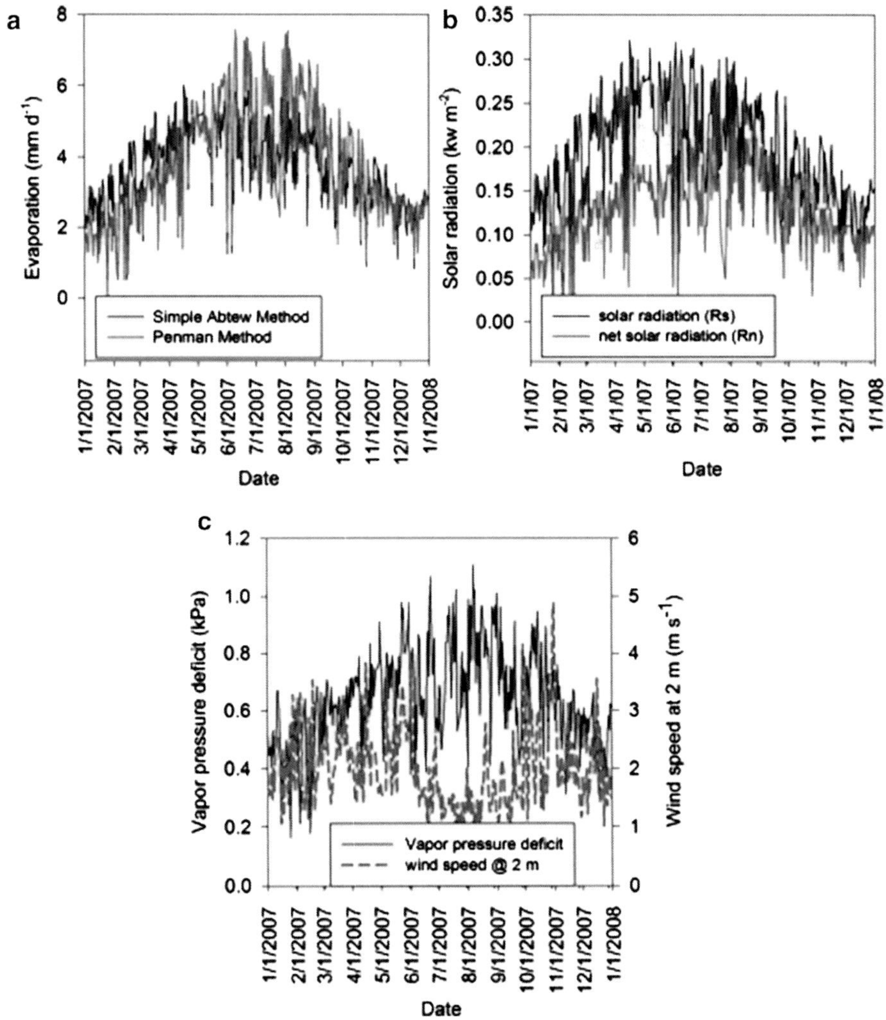

Fig. 6.12 (**a**) Penman combination method and simple Abtew methods for evaporation estimation in south Florida, (**b**) solar and net radiation, (**c**) vapor pressure deficit and wind speed

(MJ m^{-2} day^{-1}), G is heat flux (MJ m^{-2} day^{-1}), ρ is atmospheric density (kg m^{-3}), c_p is specific heat of moist air (kJ kg^{-1} °C^{-1}), $(e_\mathrm{a} - e_\mathrm{d})$ is vapor pressure deficit (kPa), r_c is canopy resistance, and r_a is aerodynamic resistance. This method has the most measured, derived, and estimated inputs, as shown in Table 6.3.

Change in heat storage (G) in soil or water is computed by Eq. 6.28:

$$G = c_\mathrm{s} d_\mathrm{s}(T_n - T_{n-1}) \tag{6.28}$$

Table 6.3 Input required for the Penman–Monteith method

Measured	Derived	Estimated
T (air temperature)	ρ (air density)	g_s (stomatal resistance)
R_n (net solar radiation)	r_c (canopy resistance)	LAI (leaf area index)
RH (relative humidity)	r_a (aerodynamic resistance)	h_c (height of cover)
u (wind speed)	e_a–e_d (vapor pressure deficit)	d (displacement height)
P (air pressure)	Δ (slope of saturation vapor pressure curve)	z_o (aerodynamic roughness height)
	γ (psychrometric constant)	z_{om} (momentum roughness height)
	G (heat storage)	c_p (heat capacity)
	λ (latent heat of vapor)	

where c_s is soil or water heat capacity (2.100 MJ m^{-3} °C^{-1} or 4.18 MJ m^{-3} °C^{-1}, respectively), d_s is effective depth (m), and T_n and T_{n-1} are average air temperature on day n and previous day.

6.3.2.4 Canopy Conductance (g_c) and Canopy Resistance (r_c)

According to Monteith (1973), water vapor loss from a leaf by diffusion is equivalent to an electrical circuit with cuticular resistances being analogous to resistance to current flow. Canopy resistance is the inverse of canopy conductance. Various authors have presented methods to estimate canopy conductance and resistance (Weert and Kamerling 1974; Slabbers 1977; Katerji and Perrier 1983; Lafleur and Rouse 1988; Allen et al. 1989; Kim and Verma 1991; Saugier and Katerji 1991; Steiner et al. 1991; Lafleur and Roulet 1992; Todorovic 1999; Katerji and Rana 2008). In FAO-P-M daily reference ET estimation method from a reference crop of known height, fixed canopy resistance (70 s m^{-1}) is recommended (Allen et al. 1998). Theoretical and empirical approaches have been used to show canopy resistance is related to soil moisture, available energy, vapor pressure deficit, and aerodynamic resistance (Gharsallah et al. 2009). Choudhury and Idso (1985) proposed that wheat stomatal conductance is a function of net solar radiation and canopy resistance is a function of leaf area index by canopy strata and canopy conductance (Eq. 6.29):

$$g_c = \sum_{j=1}^{n} L_j S_{cj} \tag{6.29}$$

where g_c is canopy conductance, L_j is leaf area index for canopy strata j, and S_{cj} is stomatal conductance of leaf strata L_j. Currently, Eq. 6.30 is widely used for estimating canopy resistance (Allen et al. 1989):

$$r_c = \frac{r_s}{0.5\mathrm{LAI}} \tag{6.30}$$

where r_c is average daily bulk canopy resistance (s m^{-1}), r_s is average minimum daytime value of stomatal resistance (s m^{-1}) for a single leaf, and LAI is leaf area index. Equations 6.29 and 6.30 are similar in form when resistance is substituted for conductance in Eq. 6.29. Canopy resistance was reported to be related to water stress ranging from 30 to 100 s^{-1} m^{-1} for the equatorial forest of Kenya (Szeicz and Long 1969).

Abtew et al. (1995) conducted experimental work by measuring stomatal conductance with a porometer on cattail plants in a lysimeter to develop a canopy resistance parameter. The design and operations of the cattail lysimeter are presented in Chaps. 3 and 7. The objective was to develop a canopy resistance parameter (r_c), measure all weather parameters needed to compute ET, measure ET with the lysimeter, and apply the P–M method with the developed r_c value. Then, compare the ET estimates to lysimeter measurements. The weather station measured solar radiation, net solar radiation, photosynthetic photon density flux (PPFD), air temperature, humidity, atmospheric pressure, and water temperature. Wind speed was measured at 1, 2.6, and 10 m for the purpose of developing the wind profile and estimating aerodynamic resistance (r_a) so that the only variable left is r_c. Wind speed was measured every 10 s and averaged every 15 min. All other parameters were measured every 5 min and averaged every 15 min. Leaf conductance and transpiration were measured with a LI-1600M steady state porometer with an aperture area of 1 cm^2.

Cattails have symmetrically arranged leaves ranging from two to four leaves on each side. Adaxial (back) leaf surfaces are concave, and abaxial (front) leaf surfaces are convex. A sampling method was used to select representative locations for measurement of conductance and transpiration from a plant. Measurements were made on the sunlit side of each plant, in the middle of the upper half (apical) and in the middle of the lower half (basal) sections of inner, middle, and outer leaves. Leaf conductance (gm), leaf temperature, photosynthetic photon flux density (PPFD), and leaf transpiration were measured between 9:45 am and 5:00 pm on April 8 and 29, 1993. Measurements were made during clear skies. A total of 208 leaf conductance measurements were made from 30 plants (Abtew et al. 1995). Figure 6.13 depicts each observation (one side of leaf) of stomatal conductance with a porometer. The following equation (Eq. 6.31) was used for computing stomatal conductance, g_s(LI-COR, Inc., 1989):

$$g_s = \frac{g_b g_m}{g_b - g_m} \tag{6.31}$$

where g_b is the boundary layer conductance inside the porometer cubicle and g_m is the measured conductance of the leaf (sum of abaxial and adaxial sides) in mol m^{-2} s^{-1}. The boundary conductance, g_m, measured in the laboratory with a wet filter was 2.26 mol m^{-2} s^{-1}. Molar conductance units were converted to velocity units based on Eq. 6.32 (LI-COR, Inc., 1989).

$$g_{sv} = \frac{8.314 g_s (T_{avg} + 273)}{P} \tag{6.32}$$

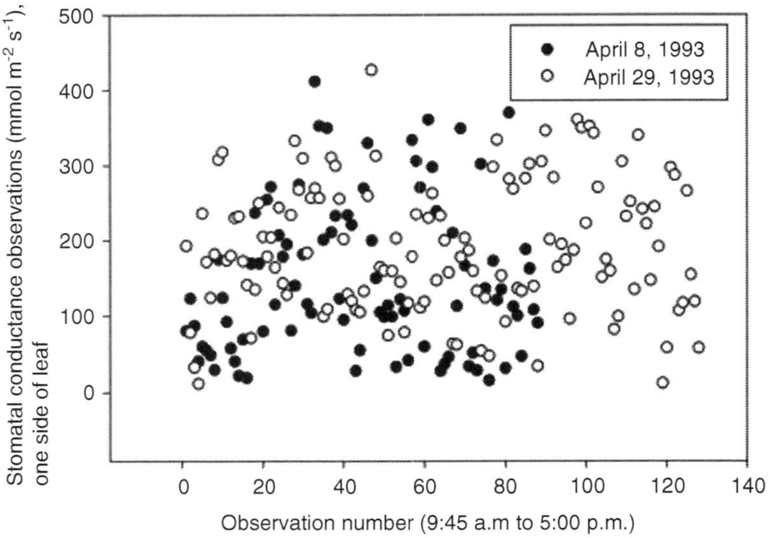

Fig. 6.13 Stomatal conductance observations over 2 days (single side of leaf)

Table 6.4 Cattail leaf conductance (g_m), stomatal conductance (g_s, g_{sv}), and other parameters

		Leaf		
	Parameter	Outer	Middle	Inner
Apical leaf	g_m (mol m^{-2} s^{-1})	0.356	0.48	0.456
	g_s (mol m^{-2} s^{-1})	0.423	0.609	0.572
	g_{sv} (mm s^{-1})	0.010	0.015	0.014
	Leaf T (°C)	28.09	27.91	27.92
	PPFD (μmol m^{-2} s^{-1})	1,885	1,889	1,721
	Transpiration (mol m^{-2} s^{-1})	7.9	10.8	10.1
	Leaf area index (m^2)	0.309	0.39	0.236
Basal leaf	g_m (mol m^{-2} s^{-1})	0.219	0.352	0.286
	g_s (mol m^{-2} s^{-1})	0.243	0.417	0.327
	g_{sv} (mm s^{-1})	0.005	0.009	0.007
	Leaf T (°C)	28	27.8	27.6
	PPFD (μmol m^{-2} s^{-1})	1,845	1,747	1,682
	Transpiration (mol m^{-2} s^{-1})	4.8	7.4	6
	Leaf area index (m^2)	0.303	0.39	0.204

Modified and adopted from Abtew et al. (1995)

where g_{sv} is stomatal conductance (mm s^{-1}), g_s is stomatal conductance (mol m^{-2} s^{-1}), 8.314 is the gas constant (Pa m^3 mol^{-1} K^{-1}), T_{avg} is average leaf temperature (°C), and P is average atmospheric pressure (Pa). Observed and computed parameters from the experiment are shown in Table 6.4 adopted from Abtew et al. (1995). The objective of the experiment was to derive canopy resistance (r_c) for cattails, a parameter needed for application of the Penman–Monteith equation from measured stomatal conductance. Canopy resistance (r_c) is the inverse of canopy conductance. Canopy conductance was derived from mean stomatal conductance as a summation of leaf area-weighted stomatal conductance of apical and basal sections of the upper, middle, and inner leaves of 30 plants (Eq. 6.33). Comparative approaches are reported in the literature (Roberts et al. 1980; Saugier and Katerji 1991).

$$g_c = \sum_j^P \sum_i^L \left((g_{svji})^l + (g_{svji})^u \right) \mathrm{LAI} \qquad (6.33)$$

where g_c is canopy conductance (mm s^{-1}); g_{svji} is mean stomatal conductance (mm s^{-1}) for adaxial leaf side (l), abaxial leaf side (u), profile (section) j, and leaf (layer) i; P is leaf profile; L is leaf layer; and LAI is leaf area index.

The leaf area-weighted and boundary layer-corrected canopy conductance (g_c) for cattails was 19.9 mm s^{-1}. Canopy resistance (r_c) is the inverse of canopy conductance (g_c) and is reported in m s^{-1} unit. Canopy resistance is derived as follows and is reported in s m^{-1} (Eq. 6.34):

$$r_c = \frac{1{,}000}{g_c} \qquad (6.34)$$

where r_c is m s^{-1} and g_c is in mm s^{-1}. The seasonal canopy resistance for cattails in south Florida, derived from Eq. 6.30, is 50.3 m s^{-1}. Comparison of the sum of squares of error between lysimeter measurements and Penman–Monteith model ET computation ($r_c = 50.3$ m s^{-1}) shows that r_c values of 40–70 m s^{-1} produce very close results.

6.3.2.5 Aerodynamic Resistance (r_a)

The aerodynamics resistance has been commonly presented as mainly a function of surface characteristics and wind speed. Equation 6.35 (Allen et al. 1989) has been in use for a while:

$$r_a = \frac{\ln \frac{(z-d)}{z_{om}}}{k^2} \times \frac{\ln \frac{(z_h-d)}{z_{oh}}}{u_z} \qquad (6.35)$$

where r_a is aerodynamic resistance (s m^{-1}), z is the height of wind measurement (m), z_h is the height of air temperature and humidity measurement (m), d is displacement height (m), z_{om} is roughness length for momentum transfer, z_{oh} is roughness length for vapor and heat transfer, k is the von Karman constant for turbulent diffusion (0.41), and u_z is wind speed measurement at height z. Application of Eq. 6.35 on a daily basis over a year in south Florida resulted in a mean r_a of 83.3 s m^{-1}. Variation of the daily mean r_a is shown in Fig. 6.14.

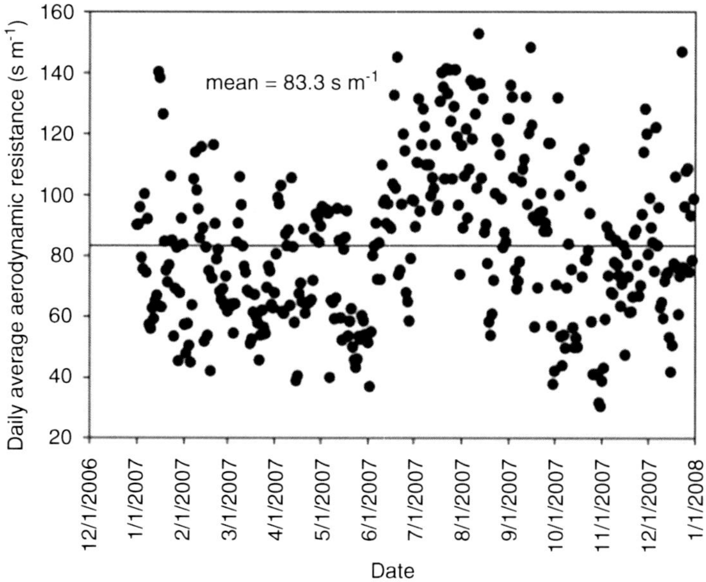

Fig. 6.14 Variation of daily mean aerodynamic resistance (r_a)

There are several equations to estimate displacement height (d) and aerodynamic roughness (z_o or z_{om}), as shown in Abtew et al. (1989). Applying the author's methods (Eqs. 6.36 and 6.37), wind speed at 2-m height is estimated (Abtew et al. 1989). Z_{oh} is estimated by Eq. 6.38 (Allen et al. 1989):

$$d = F_c h_c \tag{6.36}$$

where F_c is fraction of surface cover and h_c is average height of cover.

$$z_{om} = 0.13(h_c - d) \tag{6.37}$$

$$z_{oh} = 0.1 z_{om} \tag{6.38}$$

Application of the Penman–Monteith method is shown in Chaps. 7 and 9. Measured input of known data quality from weather stations and derived and estimated parameters from wetland surfaces were used.

6.4 Remote Sensing Methods

The latest technology of satellite-based environmental monitoring holds promise for providing meteorological variable observations for large areas. Satellite-based evapotranspiration estimation methods and applications are presented in detail in Chaps. 10, 11, and 12.

6.5 Summary

In this chapter, several ET estimation methods are presented along with input requirements. While complex methods approach physical representation of the ET process, the number of input parameters required increases. The cost of acquiring input parameters and maintaining acceptable data quality increases with the complexity of method. The virtue of application of simpler methods is well demonstrated.

References

Abtew W (1996) Evapotranspiration measurements and modeling for three wetland systems in South Florida. J Am Water Resour Assoc 127(3):140–147

Abtew W (2001) Evaporation estimation for Lake Okeechobee in South Florida. J Irrig Drain 127(3):140–147

Abtew W (2005) Evapotranspiration in the Everglades: comparison of Bowen ratio measurements and model estimates. In: Proceedings of the 2005 ASAE annual international meeting. ASAE, St. Joseph, MI

Abtew W, Obeysekera J (1995) Lysimeter study of evapotranspiration of cattails and comparison of three estimation methods. Trans ASAE 38(1):121–129

Abtew W, Gregory JM, Borrelli J (1989) Wind profile: estimation of displacement height and aerodynamic roughness. Trans ASAE 32(2):521–527

Abtew W, Newman S, Pietro K, Kosier T (1995) Canopy resistance studies of cattails. Trans ASAE 38(1):113–119

Abtew W, Obeysekera J, Iricanin N (2011) Pan evaporation and evaporation trends in south Florida. Hydrol Process 25:958–969

Allen RG, Pruitt WO (1986) Rational use of Blaney-Criddle formula. J Irrig Drain Eng 112(2)

Allen RG, Jensen ME, Wright JL, Burman RD (1989) Operational estimation of reference evapotranspiration. Agron J 81:650–662

Allen RG, Pereira LS, Raes D, Smith M (1998) Crop water requirements. FAO irrigation and drainage paper no. 56. FAO, Rome

Choudhury BJ, Idso SB (1985) An empirical model for stomatal resistance of field-grown wheat. Agric Forest Meteorol 36:65–82

Coulomb CV, Legesse D, Gasse F, Travi Y, Chernet T (2001) Lake evaporation estimates in tropical Africa (Lake Ziway of Ethiopia). J Hydrol 245:1–18

Delclaux F, Coudrain A (2005) Optimal evaporation models for simulation of large lake levels: application to Lake Titicaca, South America. Geophys Res Abstr 7:53–65

Doorenbos J, Pruitt WO (1977) Guidelines for predicting crop water requirements. FAO irrigation and drainage paper 24. FAO, Rome

Dugas WW, Fritschen LJ, Gay LW, Jeld AA, Matthias AD, Reicosky DC, Steduto P, Steiner JL (1991) Bowe ratio, eddy correlation, and portable chamber measurements of sensible and latent heat flux over irrigated spring wheat. Agric Forest Meteorol 56(1–2):1–20

Enku T, van der Tol C, Gieske ASM, Rientjes THM (2011) Evapotranspiration modeling using remote sensing and empirical models in the Fogera floodplain, Ethiopia. In: Melesse A (ed) Nile River Basin: hydrology, climate and water use. Springer, Dordercht, p 163, 170

Farnsworth RK, Thompson ES, Peck EL (1982) Evaporation atlas for the contiguous 48 United States. NOAA Technical Report NWS 33. National Weather Service, Washington, DC, June 1982

Frevert DK, Hill RW, Braaten BC (1983) Estimation of FAO evapotranspiration coefficients. J Irrig Drain ASCE 109(IR2):265–270
German ER (2000) Regional evaluation of evapotranspiration in the Everglades. USGS Water Resources Investigations Report 00–4217. USGS, Tallahassee, FL
Gharsallah O, Moncini M, Rana G (2009) Bulk canopy resistance: determination and modeling for actual evapotranspiration estimation of maize. IX Convegno Nazionale dell' Associazione Italiana di Ingegneria Agraria, Ischia Porto, 12–16 Settembre 2009, memoria n. 2–30
Hansen S (1984) Estimation of potential evapotranspiration. Nord Hydrol 15:205–212
Hargreaves GH, Samani ZA (1985) Reference crop evapotranspiration from temperature. Appl Eng Agric 1(12):96–99
Jacobs JM, Myers DA, Andersoin MC, Diak GR (2002) GOES surface insolation to estimate wetland evapotranspiration. J Hydrol 266:53–65
Jensen ME (1974) Consumptive use of water and irrigation water requirements. ASCE, New York
Katerji N, Perrier A (1983) A modélisation de l'évapotranspiration réelle d'une parcelled de luzerne: rôle d'un coefficient culltural. Agronomie 3(6):513–521 (in French)
Katerji N, Rana G (2008) Crop evapotranspiration measurement and estimation in the Mediterranean region. INRA-CRA, Bari
Kim J, Verma SB (1991) Modeling canopy stomatal conductance in a temperate grassland ecosystem. Agric Forest Meteorol 55:149–166
Lafleur PM, Roulet NT (1992) A comparison of evapotranspiration rates from two fens of the Hudson Bay Lowland. Aquat Bot 44:59–69
Lafleur PM, Rouse WR (1988) The influence of surface cover and climate on energy partitioning and evaporation in subarctic wetland. Bound Layer Meteorol 44:327–3
LI-COR, Inc (1989) LI-600 steady state porometer instruction manual. Pub. No. 8210–0030, Lincoln, Nebraska
Maidment DR (ed) (1993) Handbook of hydrology. McGraw-Hill, Inc., New York
Makkink GF (1957) Testing the Penman formula by means of lysimeters. J Inst Water Eng 11: 277–288
Melesse A, Abtew W, Dessalegne T (2009) Evaporation estimation of Rift Valley Lakes: comparison of models. Sensor J 9:9603–9615. doi:10.3390/s91209603
Monteith JL (1973) Principles of environmental physics. Edward Arnold, London
Oudin L, Hervieu F, Michel C, Perrin C, Andreassian V, Anctil F, Loumagne C (2005) Which potential evapotranspiration input for a lumped model part 2-towards a simple and efficient potential evapotranspiration model for rainfall-runoff modeling. J Hydrol 303:290–306
Penman HL (1948) Natural evaporation from open water, bare soil and grass. Proc R Soc Lond A Math Phys Sci 193(1032):120–145
Roberts J, Pymar CF, Wallace JS, Pitman RM (1980) Seasonal changes in leaf area, stomatal and canopy conductance and transpiration from bracken below a forest canopy. J Appl Ecol 17: 409–422
Saugier B, Katerji N (1991) Some plant factors controlling evapotranspiration. Agric Forest Meteorol 54:263–277
Shoemaker WB, Sumner DM (2006) Alternate corrections for estimating actual wetland evapotranspiration from potential evapotranspiration. Wetlands 26(2):528–543
Shuttleworth WJ (1993) Chapter 4: Evaporation. In: Maidment DR (ed) Hand book of hydrology. McGraw-Hill, Inc., New York
Slabbers PJ (1977) Surface roughness of crops and potential evapotranspiration. J Hydrol 34: 181–191
Steiner JL, Howell TA, Schneider AD (1991) Lysimetric evaluation of daily potential evapotranspiration models for grain sorghum. Agron J 83:240–247
Szeicz G, Long IF (1969) Surface resistance of crop canopies. Water Resour Res 83(3):622–633
Todorovic M (1999) Single-layer evapotranspiration model with variable canopy resistance. J Irrig Drain Eng 125(5):235–245
Weert RVD, Kamerling GE (1974) Evapotranspiration of water hyacinth (*Eichhornia crassipes*). J Hydrol 22:201–212

Xu CY, Singh VP (2000) Evaluation and generalization of radiation-based methods for calculating evaporation. Hydrol Process 14:339–349

Zhai L, Feng Q, Li Q, Xu C (2009) Comparison and modification of equations for calculating evapotranspiration (ET) with data from Gansu Province, Northwest China. Irrig Drain. doi:10.1002/ird.502

Chapter 7
Wetland Evapotranspiration

Abstract Wetland, marsh, bog, and fen evapotranspiration (ET) rates historically were estimated far higher than open water evaporation. Recent studies have shown that wetland evapotranspiration is not higher than open water evaporation. Lysimeter studies in south Florida show that there is no significant difference in evapotranspiration between cattails, mixed marsh, and open water. Bowen ratio evapotranspiration measurements also showed wetland evapotranspiration being not more than open water evaporation. Simple equations based on solar radiation and temperature can provide estimates of evaporation and ET in regions where most of the variation in ET is explained by one or two parameters.

Keywords Wetland evapotranspiration • Lysimeter measurements • Wetland evapotranspiration estimation methods

7.1 Introduction

Wetlands are ecosystems with open water and wetland vegetation features and periodic variation in the type and density of vegetation cover and water levels. Wetlands are subject to hydrologic variation, but mostly surface or subsurface water is available for evaporation and evapotranspiration except in regions that experience periodic severe droughts. Historically, wetlands were not of great economic interest, which might have contributed to the relatively limited study of their hydrology. Evapotranspiration is one of the major parameters of wetland hydrology. There has been lack of consensus on rates of evaporation losses from wetland features. As a major component of the hydrologic cycle, there is a need for reasonably accurate estimates of evaporation from water bodies and evapotranspiration from vegetation. Evapotranspiration depends on the availability of energy, the mechanism of mass transfer, energy transfer, and the availability of water. Evaporation and evapotranspiration are functions of solar radiation, temperature, wind speed, vapor pressure deficit, atmospheric pressure, characteristics of the surrounding environment, and

W. Abtew and A. Melesse, *Evaporation and Evapotranspiration: Measurements and Estimations*, DOI 10.1007/978-94-007-4737-1_7,

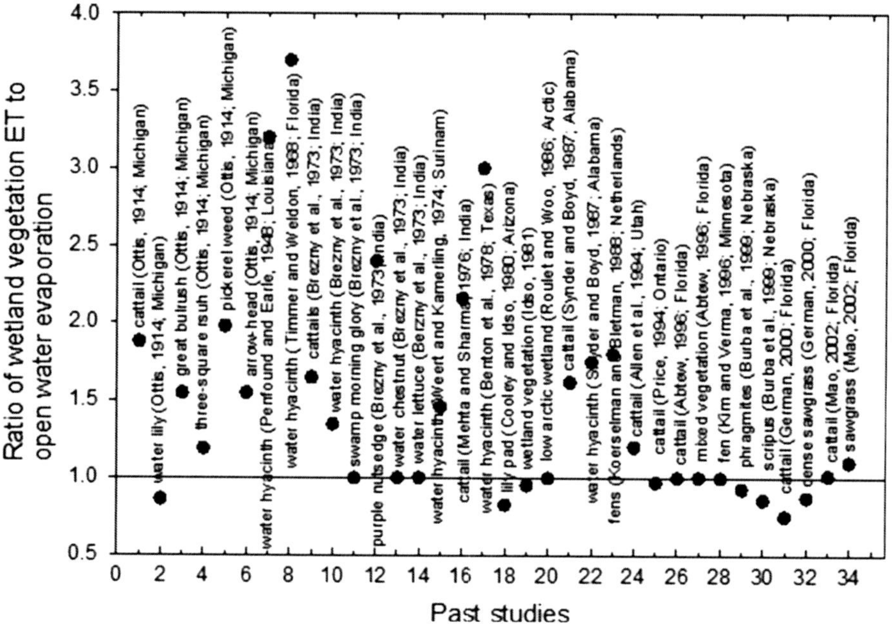

Fig. 7.1 Reported ratios of wetland evapotranspiration to open water evaporation (Abtew 2005; Abtew and Obeysekera 1995)

type and condition of vegetation. The existence of both open water and wetland vegetation in one environment has resulted in different views of what the rate of evapotranspiration could be from such systems. A shallow lake drying out due to hydrological drought could be observed invaded with vegetation, and the drying could be mistakenly attributed to an increased wetland vegetation ET.

In the past, there has been a general belief supported by small-scale experiments that wetland vegetation evapotranspiration is far higher than open water evaporation. There were cases where small pot studies were influenced by the surrounding environment. Estimated rates of wetland evapotranspiration as high as three times open water evaporation have been reported. A literature review of studies of evapotranspiration of wetland vegetation indicated that there are diverse opinions on the ratio of wetland vegetation evapotranspiration to evaporation from shallow open water surfaces. Figure 7.1 chronologically depicts various measurements and estimates of ratio of evapotranspiration from wetland vegetation to open water evaporation for many locations through the years. The reported ratios of wetland vegetation evapotranspiration to open water evaporation range from 0.75 for cattails (German 2000) in Florida to 3.7 for water hyacinth (Timmer and Weldon 1968) in Florida. Recent studies generally show the trend of reporting where wetland ET is not markedly higher or lower than shallow open water evaporation. In India, after conducting tests in 0.36 m^2 and 0.6 m deep concrete tanks, Mehta and Sharma (1976) reported a 2.16 ratio for *Typha angustata* evapotranspiration and open water evaporation. Weert and Kamerling (1974) discuss the experiment of Penfound and Earle stating that the experimental containers were placed on a laboratory balcony making clear that border effects influenced the reported rate of water hyacinth ET in Louisiana being over three times that of open water. Lafleur and Roulet (1992) studied evapotranspiration from a sedge-covered mineral-rich fen and sphagnum carpet mineral-poor fen in the southern part of the Hudson Bay in Canada. They concluded that both fen surfaces evaporate less than open water in contradiction to much of the previous literature.

Idso (1981), after reviewing literature and conducting experiments, concluded that evapotranspiration from an expansive water body does not increase measurably by the introduction of wetland vegetation. Based on experimental study in 0.6 m^2 and 0.75 m deep tanks in Fort Pierce, Florida, Debusk et al. (1983) concluded that ET rates of water hyacinth increased with plant density. They also pointed out wetland vegetation ET was correlated with open water evaporation, solar radiation, and mean daily temperature. Snyder and Boyd (1987) studied evapotranspiration of water hyacinth and *Typha latifolia* in Alabama using 5.8 m^2 and 0.41 m deep tanks. They concluded that the ratio of evapotranspiration to open water evaporation was 1.75 and 1.62 for water hyacinth and *Typha latifolia*, respectively. They remarked that evapotranspiration of Typha was highly correlated with solar radiation and leaf area index. After reviewing Snyder and Boyd's results, Idso and Anderson (1988) indicated that the high ratio of emergent macrophyte ET to open water evaporation is due to the contribution of the peripheral or side area of the experimental vegetation clump.

Actual evapotranspiration of wetlands that do not dry out can be estimated as the theoretical atmospheric demand or potential ET of wetlands (Mitsch and Gosselink 1993; Abtew et al. 2003). In dry-out conditions, roots of macrophytes will increase ET compared to no vegetation cover. Takagi et al. (1999) reported that invasion of vascular plants in a northern Japanese bog increased ET where water level was always below ground level at both test sites. Souch et al. (1998) compared measured and model-estimated evapotranspiration from disturbed (drained) and undisturbed wetland sites and concluded that there was no substantial difference between the two sites. The drained site water levels rarely dropped below the root zone.

7.2 Wetland Evapotranspiration Measurement and Modeling

7.2.1 Lysimeters

The use of constructed wetlands for storage and water quality improvements has become a developing technology. A fully automated lysimeter system was designed and installed at the Everglades Constructed Wetland Project site in south Florida (Abtew and Hardee 1993; Abtew and Obeysekera 1995). A 2-year lysimeter study of evapotranspiration in three wetland environments (cattails, mixed marsh vegetation, and open water) was conducted in the Everglades Nutrient Removal Project, a

Fig. 7.2 (**a**) Cattails, (**b**) mixed marsh, and (**c**) open water lysimeters in a multiple cell-constructed wetland (Abtew 2005, photograph provided by South Florida Water Management District)

constructed wetland in south Florida (26° 38′ N, 80° 25′ W). Two types of three fully automated lysimeters were designed to measure in situ evapotranspiration losses from three types of wetland features. One lysimeter simulated cattails (*Typha domingensis*) in cattail marsh, the second lysimeter simulated mixed marsh vegetation (spike rush, duck potato, arrowhead, maiden cane, and saw grass) in a mixed vegetation marsh, and the third simulated open water in an open water cell of the constructed wetland. Figure 7.2 depicts cattail, mixed vegetation, and open water lysimeters. The purpose of the lysimeter study was to provide ET measurements for water budget computation for the wetland and also to calibrate ET models from high-resolution meteorology data measured at the site.

The main component of each lysimeter system is a circular polyethylene tank, 3.53 m in diameter and 91 cm deep, analog depth gage, inflow and outflow pumps, flow meters, data loggers, battery, solar panel, and a complete weather station. The tank was placed on a frame at an elevation to maintain the rim of the tank a few inches above water of the surrounding wetland with fluctuating water levels. Soils from the surrounding marsh were filled in the tank to a depth of 60 cm. Cattails or mixed marsh vegetation was planted in the two respective lysimeters from the surrounding wetland. The third lysimeter was filled with water imitating the surrounding wetland. 15-min, hourly, or daily evapotranspiration (ET) was derived from the system based on Eq. 7.1:

$$\mathrm{ET} = D_t - D_{t-1} + R_\mathrm{f} + I - O \tag{7.1}$$

where D_t and D_{t-1} are depth of water level at time t and $t-1$ measured from the bottom, R_f is rainfall, I is inflow pumping, and O is outflow pumping.

The lysimeters started and stopped operating at different dates with 688 common days of observations. An average rate of 3.7 mm day^{-1} evaporation from open water, 3.5 mm day^{-1} evapotranspiration from mixed marsh, and 3.6 mm day^{-1} evapotranspiration from cattails was reported (Abtew 1996). Figure 7.3a depicts open water evaporation; Fig. 7.3b, c depicts mixed marsh and cattail evapotranspiration, respectively, from the respective lysimeters. A conclusion from the study was that there is no significant difference between evapotranspiration of wetland vegetation and evaporation from a shallow water body. The design of the lysimeters is discussed in Chap. 3.

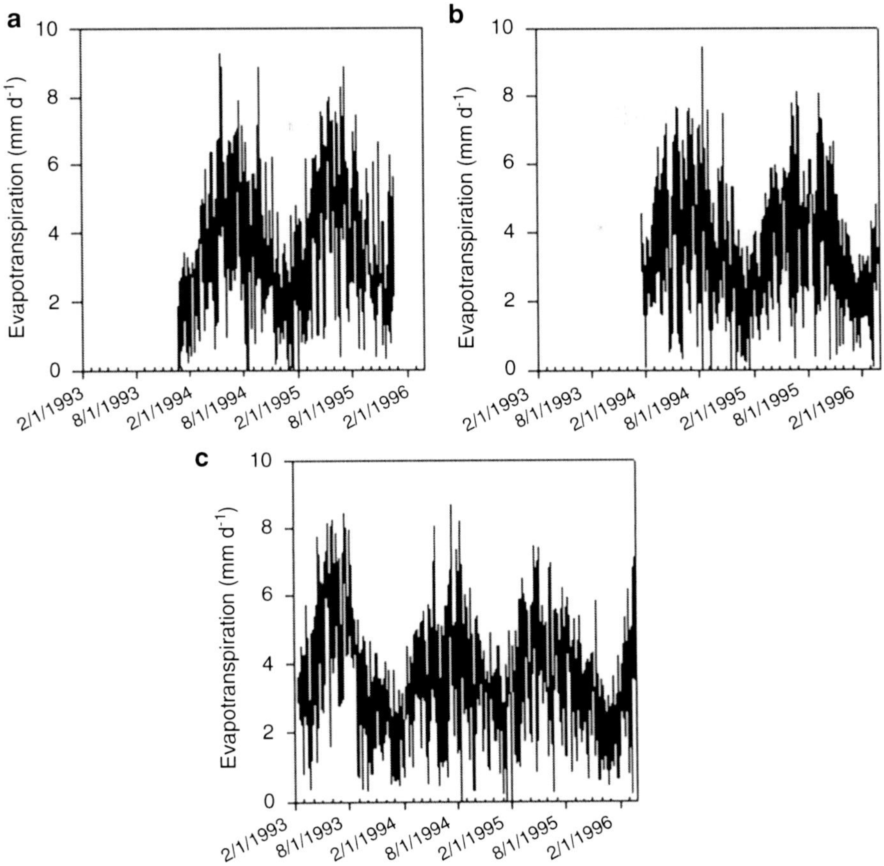

Fig. 7.3 (**a**) Daily open water evaporation, (**b**) daily mixed marsh evapotranspiration, and (**c**) daily cattails evapotranspiration

7.2.2 *Wetland ET Modeling from Lysimeter Observations*

Since the lysimeters were not designed for long-term ET monitoring, there was the need to calibrate and test ET models for long-term data acquisition. The results of the lysimeter study were applied to test and calibrate six evaporation and evapotranspiration estimation models, from simple to complex, using data acquired from weather stations at the lysimeter sites. The methods include two newly developed methods: the simple Abtew method and a radiation–temperature method. The Turc method was modified and applied by substituting daily maximum air temperature for daily average air temperature in the original equation. The Penman combination and the Penman–Monteith methods were also calibrated and applied. The simple method required a single-measured parameter and achieved comparable performance to the complex methods with numerous input requirements, as shown in Abtew (1996) and Chaps. 6 and 8.

Lysimeter-measured daily ET and weather parameters showed that ET was correlated with solar radiation ($r = 0.73$), vapor pressure deficit (0.59), minimum relative humidity ($r = 0.46$), and maximum air temperature ($r = 0.36$). Most of the variance is explained by solar radiation.

Since solar radiation explains much of the variation in wetland evapotranspiration and open water evaporation in south Florida, the potential exists to calibrate a simple solar radiation-based estimation equation. An additional advantage of solar radiation-based equations is that it eliminates the need for net solar radiation, which is more challenging to collect good quality data. Equation 7.2, which is referred to the simple Abtew method or equation, was developed from the three lysimeters' daily evapotranspiration and evaporation data and radiation measurements at the site:

$$\mathrm{ET} = K_1 \frac{R_s}{\lambda} \tag{7.2}$$

where ET is daily evapotranspiration from wetland or shallow open water or potential evapotranspiration (mm day^{-1}), R_s is solar radiation (MJ m^{-2} day^{-1}), λ is latent heat of vaporization of water (MJ kg^{-1}), and K_1 is a coefficient (0.53). The mm day^{-1} unit is derived from the fact that a kilogram of water is 1,000 cc (10^6 mm^3) and a square meter is 10^6 mm^2. Equation 7.2 estimates correlated to the average of the three lysimeters' observations with a regression coefficient of 0.7 and standard error of estimate less than 1 mm day^{-1}. The simple Abtew equation is cited, and applications in many regions are published (Abtew 1996; Xu and Singh 2000; Abtew et al. 2003; Delclaux and Coudrain 2005; Oudin et al. 2005; Shoemaker and Sumner 2006; Melesse et al. 2009; Zhai et al. 2009; Setegn et al. 2011; Enku et al. 2011).

Equation 7.3 was calibrated to estimate ET from solar radiation (R_s) and daily maximum temperature. K_3 is constant with a unit (56°C). T_{max} is daily maximum air temperature in °C. Equation 7.3 daily ET estimates correlated to the average of the three lysimeters' observations with a regression coefficient of 0.7 and standard error of estimate less than 1 mm:

$$\mathrm{ET} = \frac{1}{K_3} \frac{R_s}{\lambda} T_{\mathrm{max}} \tag{7.3}$$

Equation 7.4 is a modified Turc equation where maximum evaporation was estimated from solar radiation and air temperature. In the original Turc equation, average air temperature is used while here maximum air temperature was applied as it showed more correlation to evapotranspiration in south Florida than average air temperature. The coefficient K_2 has similar value of 0.013 as in Turc equation for computing potential evapotranspiration in a humid region:

$$\mathrm{ET_P} = K_2 \frac{(23.89 R_s + 50) T_{\mathrm{max}}}{(T_{\mathrm{max}} + 15)} \tag{7.4}$$

where ET is maximum evapotranspiration (mm day^{-1}), K_2 is a dimensionless coefficient, R_s is solar radiation (MJ m^{-2} day^{-1}), and T_{max} is maximum daily air temperature (°C). Equation 7.4 estimates correlated to the average of the three lysimeters' observations with a regression coefficient of 0.7 and standard error of estimate less than 1 mm day^{-1}.

The Priestley–Taylor equation (Eq. 7.5) is also a relatively simpler method except that it requires net solar radiation data as input. Good quality net solar radiation data acquisition requires intensive maintenance and calibration of the radiometer sensor. Experience has shown that solar radiation measurements with pyranometers are better quality than net solar radiation measurement with radiometers:

$$\mathrm{ET} = \alpha \frac{\Delta}{(\Delta + \gamma)\lambda}(R_n - G) \tag{7.5}$$

where ET is in mm day^{-1}, is slope of the vapor pressure curve (kPa °C^{-1}), γ is the psychrometric constant (kPa °C^{-1}), R_n is net radiation (MJ m^{-2} day^{-1}), and G is heat flux (MJ m^{-2} day^{-1}). The coefficient (α) in the Priestley–Taylor equation was modified from 1.26 to 1.18 to fit the model with least error of estimation and regression coefficient of 0.7 (Abtew and Obeysekera 1995).

The Penman combination equation for estimating reference evapotranspiration from grass or alfalfa in SI units is given in Eq. 7.6 (Allen et al. 1989):

$$\mathrm{ET} = \frac{1}{\lambda}\frac{\Delta(R_n - G) + \gamma 6.43(a_w + b_w u_2)(e_a - e_d)}{\Delta + \gamma} \tag{7.6}$$

where ET is in mm day^{-1}, e_a is saturation vapor pressure, e_d is actual vapor pressure, u_2 is wind speed at 2-m height in m s^{-1}, and a_w and b_w are empirical coefficients, also referred as wind coefficients, estimated as a function of day of the year. Since all other parameters in the Penman combination equation are measured or derived from measured parameters, the coefficients a_w and b_w were used as calibration coefficients to fit the model to the three lysimeters' observations with a regression coefficient of 0.7. In doing so, the regional values of the two coefficients were developed based on the normal probability density function equation applied by J.W. Wright (Allen et al. 1989). Equations 7.7 and 7.8 were calibrated and used to estimate the wind coefficients where J is day of the year (Abtew and Obeysekera 1995):

$$a_w = 0.10 + 0.2\exp\left\{-\left[\frac{J - 173}{58}\right]^2\right\} \tag{7.7}$$

$$b_w = 0.04 + 0.2\exp\left\{-\left[\frac{J - 243}{80}\right]^2\right\} \tag{7.8}$$

The performance of ET estimation models is dependent on the temporal distribution of weather parameters. Characteristic of south Florida weather is that there is sunshine due to the lower latitude and prevalent clear skies, high humidity, high temperatures, and low wind speed. Air temperature and solar radiation increase from north to south. Data from a weather station at the middle of the region is presented to display mean temporal variation of the main variables that determine the rate of evapotranspiration. Figure 7.4a depicts monthly mean of daily mean, minimum, and maximum air temperatures (1994–2010). Mean daily air temperature is 22.9°C with mean daily minimums and maximums of 18.7 and 28.2°C, respectively.

The peak months for temperature are May through October with relatively cooler temperatures from November through April. Relative humidity and wind speed are also main variables in determining the rate of ET. South Florida is a humid region with the daily maximum relative humidity averaging 96% and showing little variation from month to month. The mean and minimum relative humidity shows a pattern with the minimum in April and May. Minimum daily humidity declines from December through April and starts rising in the summer months. Figure 7.4a depicts mean monthly average air temperature (mean, minimum, and maximum). Figure 7.4b depicts mean monthly average relative humidity (mean, minimum, and maximum).

Air temperature and humidity determine saturation and actual vapor pressure. The difference between saturation and actual vapor pressure is the vapor pressure deficit which indicates available capacity of the air to hold moisture when available.

Figure 7.5a depicts solar radiation, vapor pressure deficit (×15), and wetland ET (×2). May is the peak solar radiation and peak ET month increasing from the preceding months and receding to the following months through December. An almost similar pattern is shown by vapor pressure deficit.

Generally, the region has low wind speed averaging 3.2 $\mathrm{m\,s^{-1}}$. Peak wind speed is in March with minimum wind speed in July and August. The seasonal variation of wind speed is depicted in Fig. 7.5b. In south Florida, rare events such as tropical storms as hurricanes can generate wind speed as high as 50 $\mathrm{m\,s^{-1}}$ for several hours; ET is not so important on those days as continuous rain and no sunshine conditions prevail (Abtew and Iricanin 2008). For the purpose of ET estimation, those extraordinary wind speeds need to be excluded from mean wind speed calculation to avoid bias in ET estimation.

7.2.3 Bowen Ratio–Energy Balance Method

In a U.S. Geological Survey (USGS) study, nine sites in the marshes of the Everglades in south Florida were instrumented with sensors to determine evapotranspiration from different wetland features using the Bowen ratio–energy balance method (German 2000). Field data with varying lengths of record, from 1996 to 2000, is available on the USGS web site (http://fl.water.usgs.gov/Abstracts/wri00_4217_german.html, accessed 12 December 2011). Pictures of Bowen ratio–energy

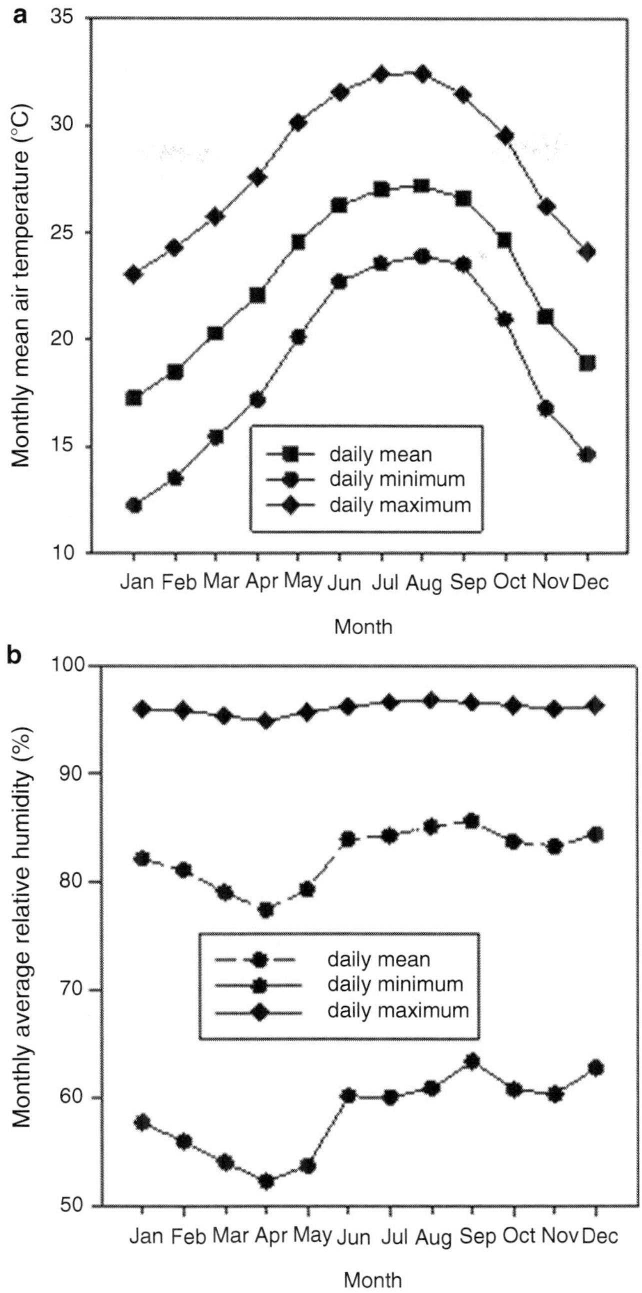

Fig. 7.4 Mean, minimum, and maximum (**a**) air temperature and (**b**) relative humidity

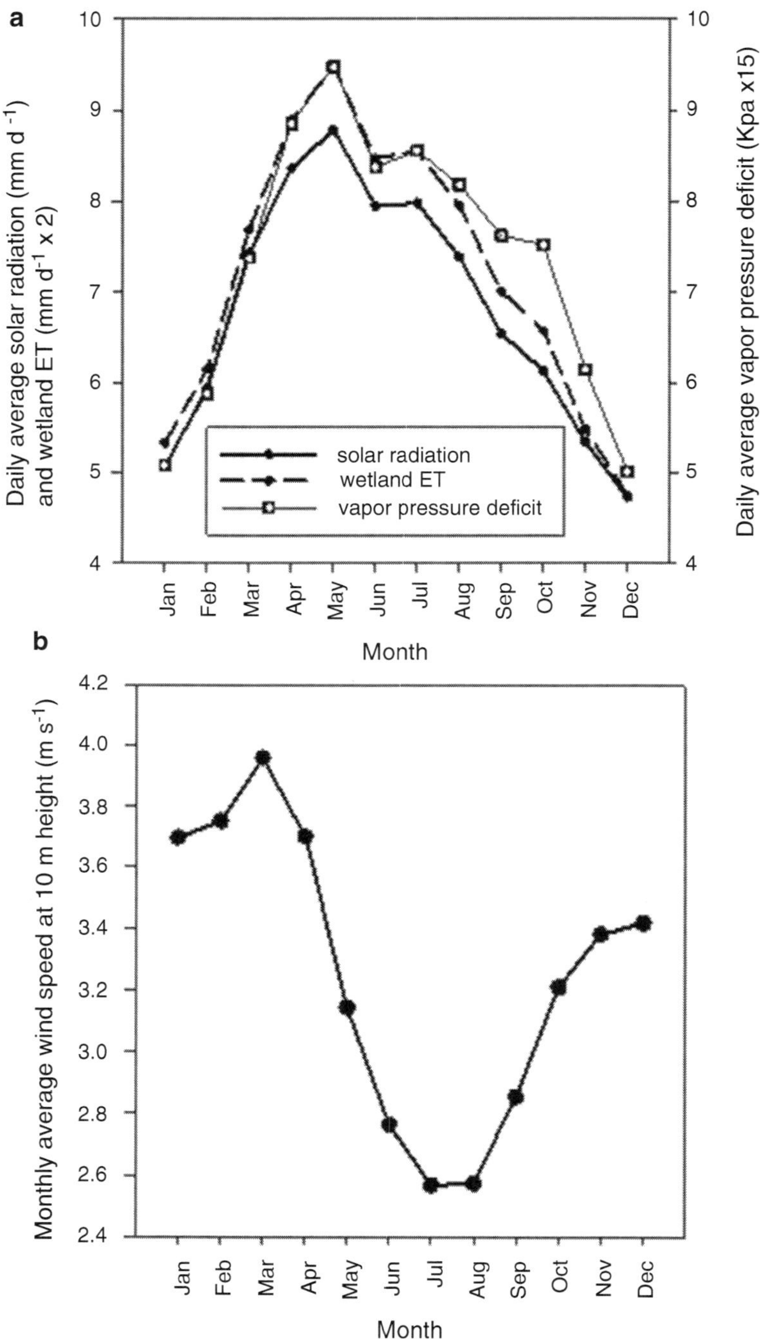

Fig. 7.5 (**a**) Solar radiation, vapor pressure deficit (×15), wetland ET (×2), and (**b**) mean wind speed

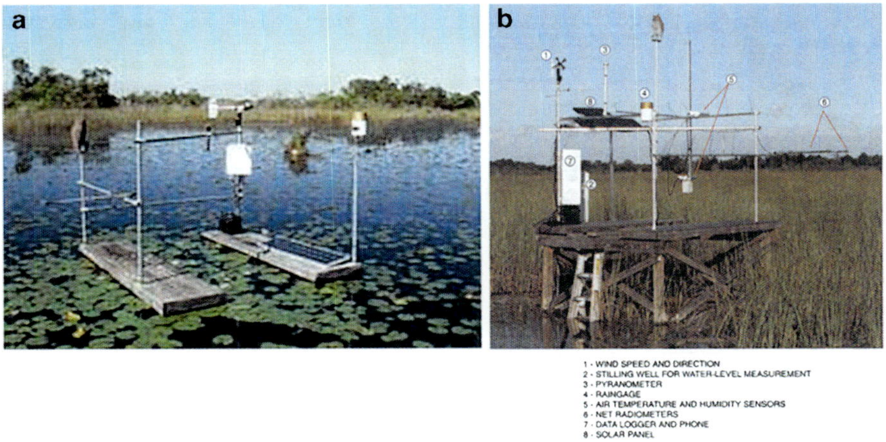

Fig. 7.6 Bowen ratio–energy balance instrumentation at (**a**) water-dominated marsh (**b**) vegetation-dominated marsh (German 2000; U.S. Geological Survey)

balance instrumentation at open water and vegetated sites are shown in Fig. 7.6 (German 2000). A location map for the sites is shown in Chap. 6. The instrumentation has net radiometer, pyranometer, wind speed and direction sensors, air temperature and humidity sensors, rain gauge, storage battery, solar panel, data logger, and cellular phone. The Bowen ratio–energy balance method is a micrometeorological method for measurement of evaporation (latent heat) with an approximate accuracy of 10% (Dugas et al. 1991). The following equation (Eq. 7.9) represents the Bowen ratio–energy balance:

$$\lambda E = \frac{R_\mathrm{n} - G}{1 + \beta} \tag{7.9}$$

where λ is latent heat of vaporization of water, E is evaporation rate, R_n is net radiation flux, G is soil heat flux, and β is Bowen ratio, which is the ratio of sensible heat (H) to latent heat (E) and derived from Eq. 7.10.

$$\beta = \frac{H}{\lambda E} = \gamma \frac{\Delta T}{\Delta e} \tag{7.10}$$

where γ is the psychrometric constant, and ΔT and Δe are finite difference of above-canopy potential temperature and vapor pressure.

The Bowen ratio instrumentation includes temperature and humidity differential with height measurements. At the Bowen ratio–energy balance ET measurement sites, sensor measurements were collected every 30 s and averaged to 15 or 30 min. Comparison of measured and model estimates of a parameter provides cross validation when the model is calibrated independently. In this case, the simple Abtew equation was calibrated with lysimeter ET measurements from a separate

Table 7.1 Comparison of Bowen ratio-measured ET and simple Abtew equation model-estimated wetland ET (Abtew 2005)

Site	No. of months	*r*	MSE mm^2	Bowen ratio-measured ET mm day^{-1}	Model-estimated ET mm day^{-1}	Site characteristics
1	24	0.90	0.20	3.36	3.54	Cattail
2	13	0.89	0.79	4.19	3.63	Open water
3	24	0.97	0.99	4.48	3.68	Open water
4	45	0.69	0.68	3.79	3.97	Dense saw grass
5	24	0.83	0.76	3.91	3.77	Medium saw grass; dry part of some years
6	32	0.80	0.50	3.63	3.80	Medium saw grass
7	58	0.82	0.99	4.19	3.97	Sparse saw grass
8	58	0.61	0.63	3.66	3.86	Sparse rushes; dry part of each year
9	24	0.70	0.72	3.40	3.89	Sparse saw grass; dry part of each year

study. Statistical comparisons of Bowen ratio–energy balance measured at each of the nine sites and the simple Abtew method-estimated average daily wetland ET for each month are presented in Table 7.1. Solar radiation data used by the simple Abtew equation was obtained from the instrumentation at each of the Bowen ratio sites, except site 2 where solar radiation data was used from a nearby weather station.

Table 7.1 presents the number of months with data (*n*), correlation coefficient (*r*), mean square error (MSE), and mean daily ET. The statistics provide a comparison between the Bowen ratio–energy balance-measured ET and the simple Abtew equation-estimated wetland ET. Site 1, the cattail marsh site, showed the smallest mean square error. The two sites with the largest difference in measured and estimated ET were sites 2 and 3. The Bowen ratio instrumentation at these open water-dominated marshes was different. While at the other seven sites, air temperature and humidity differentials were measured between two points in the air, 91–152 cm apart; at sites 2 and 3, air temperature and humidity differentials were measured 91–121 cm above the water surface. The mean estimated daily ET from all nine sites by Eq. 7.2 (3.79 mm day^{-1}) has a difference of less than 2% from the mean measured ET (3.85 mm day^{-1}) for all nine sites.

7.2.4 Penman–Monteith Method

The Penman–Monteith equation (Eq. 7.11) for evapotranspiration estimation from vegetation surfaces has numerous measured, derived, and estimated inputs, as shown in Table 6.3 and Chap. 3 (Monteith 1965):

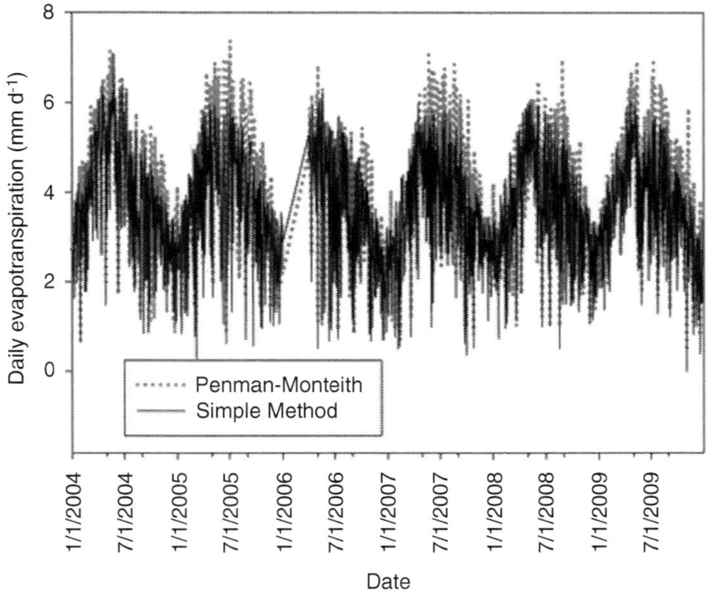

Fig. 7.7 Penman–Monteith and simple Abtew method evapotranspiration estimation from south Florida wetland

$$\mathrm{ET} = \frac{1}{\lambda}\frac{\Delta(R_\mathrm{n} - G) + \rho c_\mathrm{p}(e_\mathrm{a} - e_\mathrm{d})\frac{1}{r_\mathrm{a}}}{\Delta + \gamma\left(1 + \frac{r_\mathrm{c}}{r_\mathrm{a}}\right)} \tag{7.11}$$

where ET is in mm day^{-1}, $e_\mathrm{a} - e_\mathrm{d}$ is vapor pressure deficit in kPa, r_a is aerodynamic resistance in s m^{-1}, and r_c is canopy resistance in s m^{-1}. Details of the input in to the Penman–Monteith equation are given in Chap. 6. Application of this method to a wetland in south Florida is given as illustration of method application. The Penman–Monteith method was applied in south Florida to estimate evapotranspiration from wetlands, and the daily estimates are compared with estimates by the simple Abtew method (Fig. 7.7). The period of analysis is from January 1, 2002, to December 31, 2009, with data missing for the 3 months of January, February, and March of 2006.

The Penman–Monteith method estimates are higher for the hot and wet months of May through October with annual estimates of 1,421 mm compared to 1,335 mm for the simple Abtew method. Monthly analysis clearly displays the difference between the two methods (Fig. 7.8).

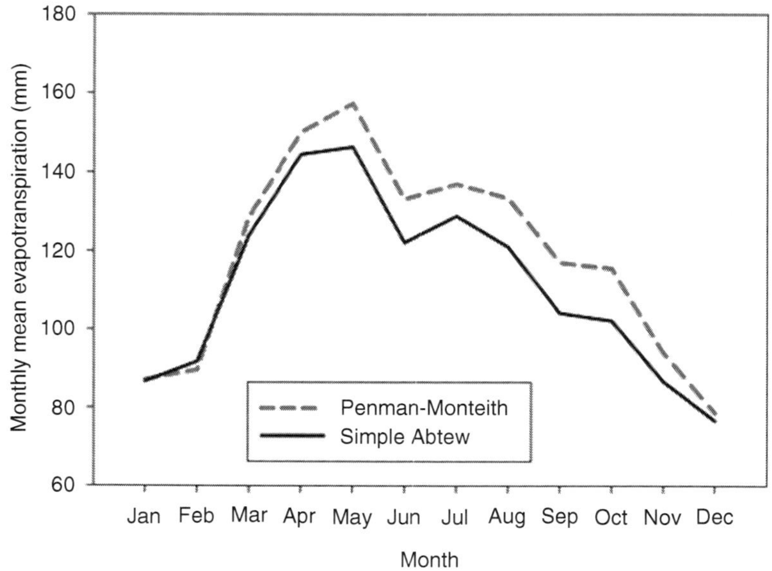

Fig. 7.8 Comparison of mean monthly evapotranspiration estimated by the Penman–Monteith and the simple Abtew method (2004–2009)

7.3 Summary

Mixed wetland vegetation and open water features of wetlands have led to many hypotheses on the rate of evapotranspiration from such features. In the past, many believed the rates are far higher than open water evaporation. Recent studies have shown that wetland evapotranspiration in many regions is not that much different from open water evaporation. The rate of evapotranspiration is controlled not only by the availability of water and the presence of vegetation but also by the availability of energy, by capacity of the air to hold moisture, and by rates of energy and mass transfer. In south Florida and many regions, simple models based on solar radiation and temperature could provide low-cost wetland evapotranspiration, open water evaporation, and potential evapotranspiration estimates. Detail of application of complex methods is presented in Chap. 6. Remote sensing applications for evapotranspiration estimation are presented in Chaps. 10, 11, and 12.

Acknowledgements We would like to acknowledge Ed German from U.S. Geological Survey for taking the photographs shown in Fig. 7.6a, b.

References

Abtew W (1996) Evapotranspiration measurements and modeling for three wetland systems in South Florida. J Am Water Resour Assoc 127(3):140–147

Abtew W (2005) Evapotranspiration in the Everglades: comparison of Bowen ratio measurements and model estimates. In: Proceedings of the 2005 ASAE annual international meeting. ASAE, St. Joseph, MI

Abtew W, Hardee J (1993) Design of a lysimeter for a wetland environment: evapotranspiration of cattails (*Typha domingensis*). Paper presented at the 1993 ASAE paper no. 93–2553. ASAE, St. Joseph, MI

Abtew W, Iricanin N (2008) Hurricane effects on south Florida water management system: a case study of Hurricane Wilma of October 2005. J Spat Hydrol 8(1):1–21

Abtew W, Obeysekera J (1995) Lysimeter study of evapotranspiration of cattails and comparison of three estimation methods. Trans ASAE 38(1):121–129

Abtew W, Obeysekera J, Ortiz MI, Lyons D, Reardon A (2003) Evapotranspiration estimation for South Florida. In: Bizier P, DeBarry P (eds) Proceedings of the world water and environmental congress 2003. ASCE, Reston

Allen RG, Jensen ME, Wright GL, Burman RD (1989) Operational estimates of reference evapotranspiration. Agron J 81:650–662

Allen RG, Hill W, Srikanth V (1994) Evapotranspiration parameters for variably sized wetlands. ASAE paper no. 93–2516. ASAE, St. Joseph, MI

Benton AR, James WP, Rouse JW Jr (1978) Evapotranspiration from water hyacinth (*Eichhornia crassipes* (Mart.) Solms) in Texas reservoirs. Water Resour Bull 14(4):919–930

Brenzy O, Mehta I, Sharma RK (1973) Studies on transpiration of some aquatic weeds. Weed Sci 21(May):197–204

Burba GG, Verma SB (1999) A comparative study of surface energy fluxes of three communities (*Phragmites australis, Scirpus acutus*, and open water) in a prairie wetland ecosystem. Wetlands 19(2):451–457

Cooley KR, Idso SB (1980) Effects of lily pads on evaporation. Water Resour Res 16(3):605–606

DeBusk TA, Ryther JH, Williams LD (1983) Evapotranspiration of *Eichhornia crassipes* (Mart.) solms and *Lemna minor L.* in central Florida: relation to canopy structure and season. Aquat Bot 1(16):31–39

Delclaux F, Coudrain A (2005) Optimal evaporation models for simulation of large lake levels: application to Lake Titicaca, South America. Geophys Res Abstr 7:53–65

Dugas WA, Fritschen LJ, Gay LW, Held AA, Mathias AD, Reicosky DC, Stedoto P, Steiner JL (1991) Bowen ratio, eddy correlation, and portable chamber measurements of sensible and latent heat flux over irrigated spring wheat. Agric Forest Meteorol 56:1–20

Enku T, van der Tol C, Gieske ASM, Rientjes THM (2011) Evapotranspiration modeling using remote sensing and empirical models in the Fogera floodplain, Ethiopia. In: Melesse A (ed) Nile River Basin: hydrology, climate and water use. Springer, Dordrecht, pp 163–170

German ER (2000) Regional evaluation of evapotranspiration in the Everglades. USGS Water Resources Investigations Report 00–4217. USGS, Tallahassee, FL

Idso SB (1981) Relative rates of evaporative water losses from open and vegetation covered water bodies. Water Resour Bull 17(1):6–48

Idso SB, Anderson MG (1988) A comparison of two recent studies of transpirational water loss from emergent aquatic machrophytes. Aquat Bot 31:191–195

Kim J, Verma SB (1996) Surface exchange of water vapour between an open sphagnum fen and the atmosphere. Bound Layer Meteorol 79:243–264

Koerselman W, Beltman B (1988) Evapotranspiration from fens in relation to Penman's potential free water evaporation (Eo) and pan evaporation. Aquat Bot 31(3–4):307–320

Lafleur PM, Roulet NT (1992) A comparison of evaporation rates from two fens on the Hudson Bay Lowland. Aquat Bot 44:59–69

Mao LM, Bergman MJ, Tai C (2002) Evapotranspiration measurement and estimation of three wetland environments in the Upper St. John's River Basin, Florida. J Am Water Resour Assoc 5(38):1271–1285

Mehta I, Sharma RK (1976) A note on water loss from Typha (Aira) weed. Ann Arid Zone 15(1, 2): 114–116

Melesse A, Abtew W, Dessalegne T (2009) Evaporation estimation of Rift Valley Lakes: comparison of models. Sensor J 9:9603–9615. doi:10.3390/s91209603

Mitsch WJ, Gosselink JG (1993) Wetlands, 2nd edn. Van Nostrand Reinhold, New York

Monteith JL (1965) Evaporation and the environment. In: The state and movement of water in living organisms, XIXth symposium of the Society of Experimental Biologists Swansea. Cambridge University Press, Cambridge

Ottis CH (1914) The transpiration of emerged water plants: its measurement and its relationships. Bot Gaz LVIII:457–494

Oudin L, Hervieu F, Michel C, Perrin C, Andreassian V, Anctil F, Loumagne C (2005) Which potential evapotranspiration input for a lumped model part 2-towards a simple and efficient potential evapotranspiration model for rainfall-runoff modeling. J Hydrol 303:290–306

Penfound WM, Earle TT (1948) The biology of the water hyacinth. Ecol Monogr 14(4):448–472

Price JS (1994) Evapotranspiration from lakeshore *Typha* marsh on Lake Ontario. Aquat Bot 48:262–272

Roulet NT, Woo M-K (1986) Wetland and lake evaporation in the low Arctic. Arct Alp Res 18(2):195–200

Setegn SG, Chowdary VM, Mal BC, Yohannes F, Kono Y (2011) Water balance study and irrigation strategies for sustainable management of a tropical Ethiopian lake: a case study of Lake Alemaya. Water Resour Manage 25(9):2081–2107. doi:10.1007/s11269-011-9797-y

Shoemaker WB, Sumner DM (2006) Alternate corrections for estimating actual wetland evapotranspiration from potential evapotranspiration. Wetlands 26(2):528–543

Snyder RL, Boyd CE (1987) Evapotranspiration by *Eichhornia crassipes* (Mart.) and *Typha latifolia* L. Aquat Bot 27:217–227

Souch C, Grimmond CSB, Wolfe CP (1998) Evapotranspiration rates from wetlands with different disturbance histories: Indiana Dunes National Lake Shore. Wetlands 18(2):216–229

Takagi K, Tsuboya T, Takahashi H, Inuoe T (1999) Effects of the invasion of vascular plants on heat and water balance in the Sarobetsu Mire, northern Japan. Wetlands 19(1):246–254

Timmer CE, Weldon LW (1968) Evapotranspiration and pollution of water by water hyacinth. Hyacinth Control J 6:4–37

Weert RV, Kamerling GE (1974) Evapotranspiration of water hyacinth (*Eichhornia crassipes*). J Hydrol 22:201–212

Xu CY, Singh VP (2000) Evaluation and generalization of radiation-based methods for calculating evaporation. Hydrol Process 14:339–349

Zhai L, Feng Q, Li Q, Xu C (2009) Comparison and modification of equations for calculating evapotranspiration (ET) with data from Gansu Province, Northwest China. Irrig Drain 58: 1–14. doi:10.1002/ird.502

Chapter 8
Lake Evaporation

Abstract Evaporation from lakes depends on available energy, mechanism of momentum, mass, and energy transfers. The surface area of the lake determines the effect of advective energy from the surrounding environment. The depth of the lake is critically important for heat storage and release. Various approaches have been used to develop lake evaporation equations that address the factors that affect the rate. In this chapter, the evaporation process and methods of estimation are presented in detail. Actual measurements of evaporation are used to calibrate models from the simplest to the most complex. The pan method, water budget method, energy balance methods, mass transfer methods, Bowen ratio method, Penman method, and radiation-based and radiation–temperature-based methods are presented in detail.

Keywords Lake evaporation • Pan evaporation • Energy balance • Evaporation estimation methods

8.1 Introduction

Lakes and reservoirs are sources of water supply, fishing for food supply, recreation, hydropower, and transportation and maintain wetland and aquatic ecosystems. Lakes' and reservoirs' evaporation rates are of great interest for water resources management. A major cause of decline in water level for lakes and reservoirs is evaporation. With ever increasing water demand, measuring or estimating evaporation rates is important. Especially in arid and semiarid areas, evaporation is a major flux in the hydrologic cycle. Lake evaporation direct measurements are generally not easy nor are reliable data available. In most cases, lake evaporation is estimated using pan evaporation, water budget, or from empirical models that mostly use data collected on land. Evaporation from lakes depends on the availability of energy and the mechanisms of mass and energy transfer. The surrounding environment of the lake, the in-lake environment, depth, and surface area of the lake affect the rate of evaporation. The effect of depth of lake on evaporation has been discussed.

W. Abtew and A. Melesse, *Evaporation and Evapotranspiration: Measurements and Estimations*, DOI 10.1007/978-94-007-4737-1_8,

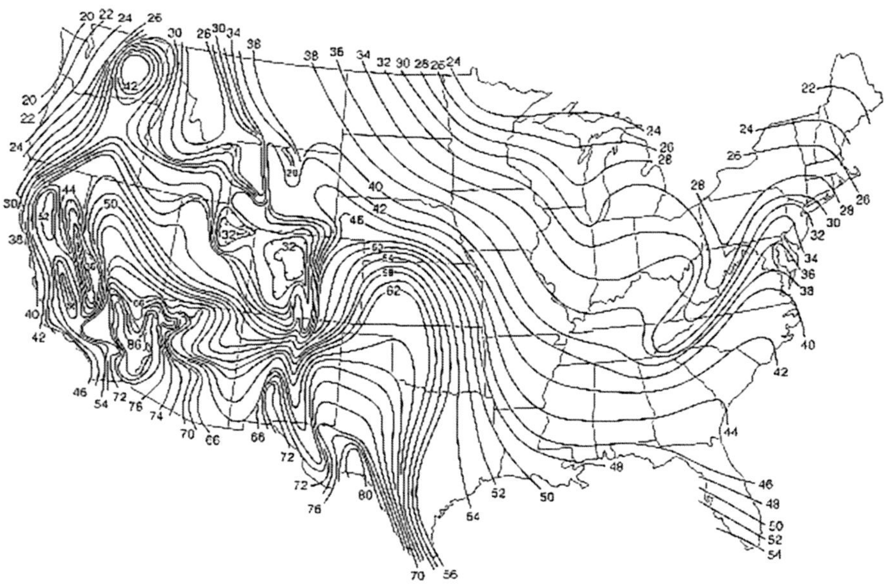

Fig. 8.1 Mean annual lake evaporation in inches over the USA (Kohler et al. 1959; U.S. Weather Bureau); 1 in. = 2.54 cm

Lake depth is suggested to be more important at higher latitudes where there is a more marked difference between summer and winter temperatures than in the tropics (Reis and Dias 1998). The importance of the vertical temperature profile and subsurface heat storage changes and depths of various lakes are presented stating that deeper depths reduce annual lake evaporation. Evaporation is a function of solar radiation, temperature, wind speed, vapor pressure deficit, atmospheric pressure, and advective and storage energy sources and sinks. Most of these variables are latitude and altitude dependent.

Annual lake evaporation in the United States varies from 51 cm in the northeast and 218 cm in southern California. A U.S. Weather Bureau 1959 technical paper provides a map, Fig. 8.1, with mean annual lake evaporation for the contiguous United States (Kohler et al. 1959). Generally, the south and southwest have higher evaporation than the north and northeast. Roberts and Stall (1967) produced a lake evaporation isohyetal map for Illinois showing a general increasing trend from the northeast to the southeast ranging from 76 to 97 cm. Based on lysimeter measurements and literature review, isohyetal lines for lake evaporation in south Florida are shown in Fig. 8.2 (Abtew et al. 2003).

Lake Okeechobee is located in the central region of south Florida at 26° 39′ and 80° 37′ longitude. The lake has a surface area of 1,732 km^2 and an average depth of 2.7 m (Jin et al. 1998). Based on 5 years of meteorological data applications in evaporation models and water budget analysis, annual average evaporation of 132 cm was reported for Lake Okeechobee (Abtew 2001). This estimate is close

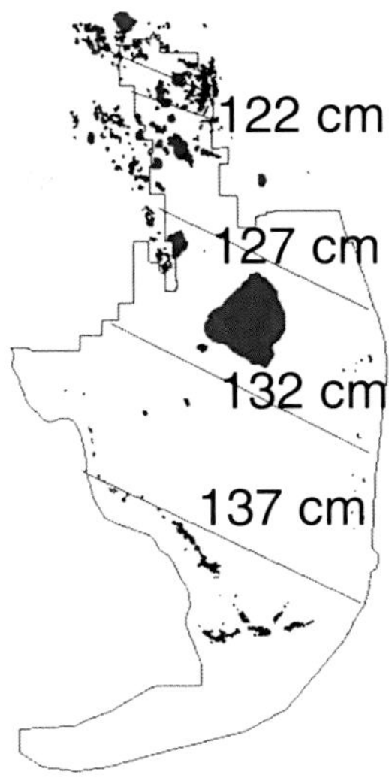

Fig. 8.2 Estimated isohyetal lines for open water evaporation, wetland evapotranspiration, and potential evapotranspiration for south Florida (Abtew et al. 2003)

to what is presented in Figs. 8.1 and 8.2. In a study on a temperate lake, Lake Sparkling in northern Wisconsin, evaporation and energy balance are presented by Lenters et al. (2005). It was reported that humidity (32%) followed by the difference in water and air temperature (22%) explains most of the interseasonal variation in lake evaporation. The results of this study show the differences in subtropical and tropical lakes' energy balances and evaporation variation. It is also reported that wind and interseasonal lake evaporation show little correspondence in that climate and mass transfer evaporation estimations were less accurate due to the inclusion of wind speed.

8.2 Lake Evaporation Estimation Methods

8.2.1 Pan Method

The most common lake evaporation estimation method is the pan method where evaporation from a small pan is related to evaporation from a lake through a pan coefficient. Various types of pans are used in different parts of the world. A common pan is the class A evaporation pan of the National Weather Service in the United States. The pan is 120.7 cm in diameter and 25 cm in depth. Water is added or removed to maintain water level at 5 cm from the rim. The pan is usually accompanied with a rain gauge to factor out the contribution of rainfall to the depth of water in the pan. The sunken Colorado pan is square in shape (100 cm × 100 cm), 50 cm deep, and buried in the ground to a depth of 45 cm. Variations between pans include setup, pan environment, measurement errors, and differences in operations

(Abtew et al. 2011). Details on evaporation pans are presented in Chap. 3. The process of acquiring evaporation estimates from a pan can be presented with a mass balance equation, Eq. 8.1:

$$E_{\text{pan}} = D_{t-1} - D_t + R_{\text{f}} - L \pm e \tag{8.1}$$

where D_t is current day depth of water in the pan and D_{t-1} is previous day depth of water measured from the top, R_{f} is rainfall, L is other losses such as bird or animal consumption, and e is errors. Sources of error in monitoring evaporation with an open outdoor pan include environmental factors such as location, wind flow obstruction, advective heat sources or losses in the area surrounding the pan, height of pan, bird guard, rate of windblown sediment accumulation, and frequency of cleanup, reading and measurement errors, and recording errors. Some pans have bird guards of meshed wire cover to deter birds from drinking or bathing. Bird guard was acknowledged for lowering evaporation rates. In an Australian case, a correction factor (7%) has been applied to correct for the effect of bird guard (Gifford et al. 2007).

Lake evaporation is estimated from pan evaporation based on Eq. 8.2, where E_{L} is lake evaporation, E_{pan} is pan evaporation, and K_{p} is pan coefficient:

$$E_{\text{L}} = E_{\text{pan}} \times K_{\text{p}} \tag{8.2}$$

Pan coefficients vary from area to area for multiple reasons. Geographical location is a factor in pan coefficient. Other reasons are variations in pan type, pan setup, pan environment, pan operator's skills, pan maintenance, rainfall measurement, losses, and errors. Wide ranges of pan coefficients have been reported. Abtew (2001) evaluated pan coefficients for Lake Okeechobee in south Florida from seven pan stations around the lake and vicinity areas. An average coefficient of 0.76 was produced from all pan stations with varying coefficients of a low of 0.64 to a high of 0.95 on an annual basis. Boyd (1985) after 1 year of fully controlled experiment in Auburn, Alabama, reported a range of pond to pan evaporation coefficients (0.72–0.90) with an average of 0.81. Morton (1986) applied the CRLE (Complementary Relationship Lake Evaporation) model for 16 lakes in North America and one lake in East Africa. Extracting pan coefficients from the reported pan evaporation and the CRLE lake evaporation estimates result in a range of pan coefficients for the 17 lakes (Table 8.1). Coefficients range from 0.59 to 0.84 with a mean of 0.69. Due to variation in thermal inertia between a lake and a pan, pan coefficients incur error in evaporation estimation for shorter periods as less than a season (Webb 1966).

8.2.2 *Water Balance Method*

Water balance is one of the simplest methods of lake evaporation estimation. As long as there are data on surface water inflows (I), outflows (O), rainfall (R_{f}), water level, and storage (S), evaporation from a lake (E_{L}) can be estimated based on Eq. 8.3:

Table 8.1 Pan coefficients (K_p) derived from published pan and lake evaporation (mm year^{-1})

Lake	E_{pan} (Morton 1986)	Lake evaporation (CRLE, Morton 1986)	K_p
Dauphin, Manitoba, Canada	859	665	0.77
Last Mountain Lake, Saskatchewan, Canada	1,005	695	0.69
Lake Ontario, North America	913	709	0.78
Utah Lake, Utah	1,945	1,235	0.63
Lake Winnemucca, Nevada	2,076	1,319	0.64
Pyramid Lake, Nevada	2,123	1,249	0.59
Lake Hefner, Oklahoma	1,778	1,286	0.72
Silver Lake, California	2,631	1,920	0.73
Salton Sea, California	3,006	1,765	0.59
Lake Victoria, East Africa	1,940	1,624	0.84
Lake Superior, North America	801	528	0.66
Great Salt Lake, Utah	1,438	1,005	0.70
Walker Lake, Nevada	1,930	1,277	0.66
Tulare Lake, California	2,237	1,464	0.65
Buena Vista Lake, California	2,535	1,535	0.61
Elsinore Lake, California	1,800	1,348	0.75
Lake Okeechobee, Florida	2,070	1,624	0.78

$$E_L = I + R_f - O - \Delta S \pm e \pm S_p \tag{8.3}$$

where ΔS is change in storage, e is errors, and S_p is seepage or groundwater movement into or out of the lake. The application of this method is limited by the completeness and quality of both surface and subsurface inflows and outflows into the lake. Rainfall over the lake surface area is usually estimated from nearby rain gauges. The number of gauges and the location of gauges are factors that affect rainfall estimation. Storage in a lake is computed using stage–storage relationships. Based on bathymetry surveys and stage–storage curves or tables, the volume of water in a lake can be estimated from the water level or water surface elevation readings. Stage–storage relationship for Lake Okeechobee in south Florida is shown in Fig. 8.3. Langbein (1951) estimated Lake Okeechobee evaporation for the 1941 through 1947 period using water budget analysis. The annual evaporation estimate of 132.5 cm is comparable to the model estimate of 132 cm (Abtew 2001). Morton (1986) presented water budget evaporation estimate for the 17 lakes cited in Table 8.1 including Lake Okeechobee. Estimates for Lake Okeechobee were higher, 156.7 cm. Mostly, the results are comparable to the CRLE model estimates.

8.2.3 Energy Balance

Energy balance is a method that is applicable to estimate evaporation from lakes. As the water balance method accounts for inflows, outflows, and change in storage

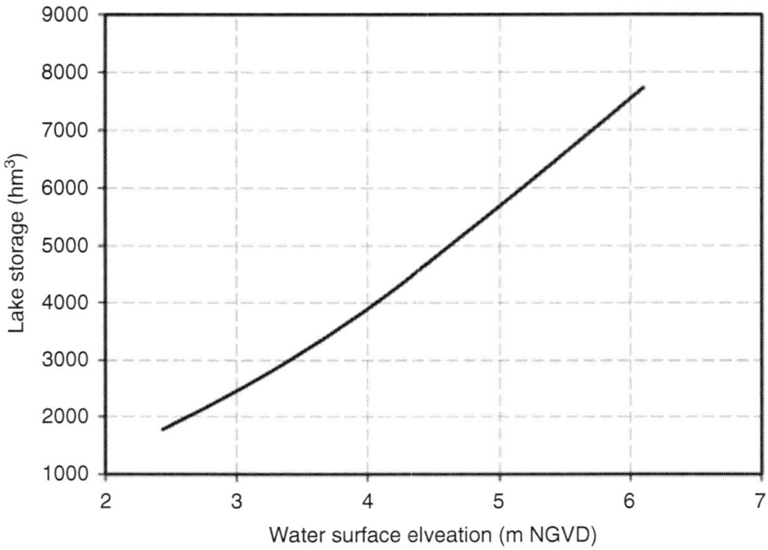

Fig. 8.3 Stage–storage curve for Lake Okeechobee

of water mass, the energy balance method accounts for energy input (Q_{in}) into the lake, energy leaving the lake (Q_{out}), and change in energy storage (ΔQ_s) in the lake. The general equation for energy balance is shown in Eq. 8.4a with error term e. Errors are from measurements of each mass balance component and completeness of source and sink accounting in the energy balance equation:

$$Q_{in} - Q_{out} = \Delta Q_s \pm e \tag{8.4a}$$

Energy inflows into the lake water body are shown in Eq. 8.4b:

$$Q_{in} = Q_{R_n} + Q_a + \Delta Q_s + Q_e + Q_h \quad \text{where } (Q_a, \Delta Q_s > 0 \text{ and } Q_e \text{ and } Q_h) \tag{8.4b}$$

where Q_{R_n} is net solar radiation, Q_a is positive net advective energy input, ΔQ_s is positive change in energy storage, Q_e is energy released from condensation at the surface of the lake, and Q_h is sensible heat lost by air at the surface of the lake. Net advective energy is energy input as a balance of energy inflows and outflows associated with surface and groundwater movement into and out of the lake. Energy outflows from the lake are shown in Eq. 8.4c:

$$Q_{out} = Q_e + Q_h + Q_a + \Delta Q_s \quad \text{where } (Q_a, \Delta Q_s < 0 \text{ and } Q_e > 0) \tag{8.4c}$$

where Q_e is energy used for evaporation, Q_h is sensible heat gained by air at the surface of the lake, Q_a is advective energy loss, and ΔQ_s is loss of stored energy. Energy lost by evaporation can be computed by Eq. 8.4d:

$$Q_e = Q_{R_n} - Q_h - Q_a - \Delta Q_s \tag{8.4d}$$

Net solar radiation is measured with instrumentation or estimated from incoming solar radiation. Energy balance of lakes is dependent on season, latitude, altitude, lake depth, surface area of lake, and surrounding environment. Salinity or dissolved solids concentrations are also cited to be a factor (Morton 1986). Winter and Rosenberry (1995) pointed out that one of the contentious issues in energy balance is the time interval of energy balance computation and the sampling of stored energy in the lake. There is uncertainty in the amount of heat stored in the lake at a certain time. Continuous measurement of representative temperature profile will aid in reducing uncertainty in heat storage in the lake. The uncertainty in estimating energy storage in the whole lake could be reduced by monitoring water temperature at a few centimeters depth to compute the energy flux into the air at the surface or into the water surface. The vertical energy balance at the surface of the lake water can be expressed by Eq. 8.5 dropping the advective energy term:

$$\lambda E = R_n - H - G \tag{8.5}$$

where λE is latent heat flux, H is sensible heat flux, and G is heat gained or lost by the upper layer of the lake. λ is latent heat of vaporization of water (Eq. 8.6):

$$\lambda = 2.501 - 0.002361 T_s \tag{8.6}$$

where T_s is water temperature in °C at lake surface and λ is in MJ kg^{-1}. Net solar radiation (R_n) is measured using hemispherical net radiometers or estimated from solar radiation measurements (Jensen 1974), as shown in Eq. 8.7:

$$R_n = (1 - \alpha) R_s - R_b \tag{8.7}$$

where α is shortwave reflectance or albedo, R_s is solar radiation, and R_b is net back or outgoing thermal radiation. If measured solar radiation data are not available, there are formulas to estimate it from clear day solar radiation or extraterrestrial solar radiation (Jensen 1974; Linacre 1993; Allen et al. 2005). Figure 8.4 depicts extraterrestrial radiation (R_a), clear sky or cloudless solar radiation (R_{so}), incoming solar radiation (R_s), and net shortwave radiation (R_n) at south Florida. The source of data for R_a at 26° latitude is FAO 1977. Cloudless solar radiation (R_{so}) was estimated as 71% of R_a. Solar radiation (R_s) was measured over Lake Okeechobee in south Florida, and net solar radiation (R_n) was measured on land close to the lake (latitude 26° 39′ and longitude 80° 37′), averaged from 2001 to 2009.

To directly apply the energy balance equation (Eq. 8.5), the estimation of sensible heat (H), heat gained or lost by air at the lake surface, is difficult. Temperature

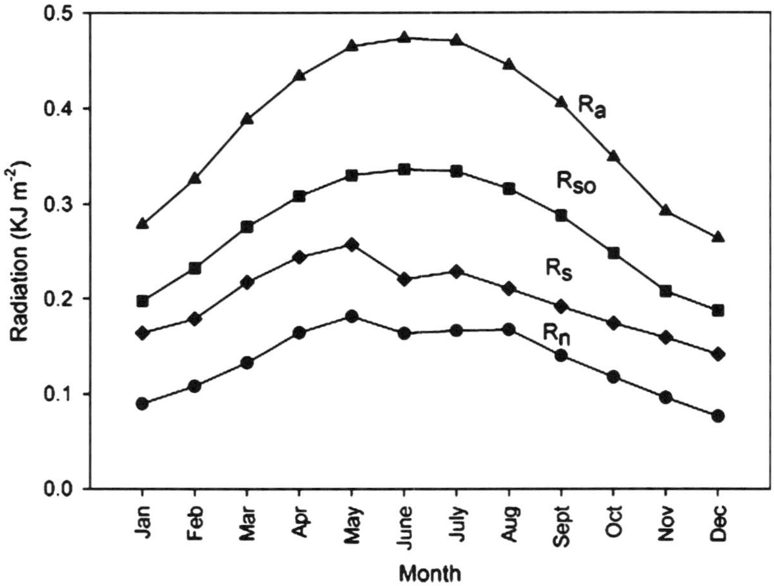

Fig. 8.4 Extraterrestrial (R_a), cloudless solar radiation (R_{so}), solar radiation (R_s), and net solar radiation (R_n) over south Florida

gradient and sensible heat transfer coefficient estimations are challenging. In modeling the evaporation and condensation processes, momentum, mass, and energy transfer mechanisms have to be accounted. The sensible heat (H) flux, shear stress (τ), and latent heat (λE) flux are presented in general form by Eqs. 8.8, 8.9, and 8.10:

$$H = \rho c_p k_h \frac{dT}{dz} \tag{8.8}$$

where ρ is air density, c_p is specific heat of air, k_h is sensible heat transfer coefficient, and dT/dz is change in temperature with height.

$$\tau = \rho k_m \frac{du}{dz} \tag{8.9}$$

where τ is shear stress, ρ is air density, k_m is transfer coefficient for shear stress, and du/dz represents the change in wind speed with height.

$$\lambda E = \frac{\lambda \varepsilon}{P} k_w \frac{de}{dz} \tag{8.10}$$

where λ is latent heat of vaporization, ε is the ratio of molecular weights of water to dry air, P is atmospheric pressure, k_w is coefficient for latent heat transfer, and de/dz is vapor pressure change with height.

The three transfer coefficients (k_m, k_w, k_h) are dependent on wind speed, vapor pressure, and temperature gradient with height. Surface conditions and atmospheric stability are also factors to be considered (Katul and Parlange 1992). For most applications, the three transfer coefficients are assumed to be equal (Federer 1970). The heat transfer coefficient (k_h) has been expressed in implicit and explicit forms. Explicit forms from various sources are presented by Eqs. 8.11, 8.12, and 8.13:

$$k_h = u_*^2 \frac{dz}{du} \tag{8.11}$$

where u_* is friction velocity and dz/du is the inverse of wind speed gradient (Monteith 1973).

$$k_h = \frac{k u_* (z - d + z_h)}{\Phi_h} \tag{8.12}$$

where k is the von Karman constant (0.41), z is height, d is displacement height, z_h is roughness length for heat transfer, and Φ_h is a stability correction factor, a function of the Monin–Obukhov length (Stannard 1993).

$$k_h = u_* \theta_* \frac{dz}{dT} \tag{8.13}$$

where θ_* is temperature scale and is computed by Eq. 8.14 as the inverse of the temperature gradient (Jacovides et al. 1992).

$$\theta_* = \frac{\Delta T k}{\ln\left(\frac{z_2}{z_1}\right)} \tag{8.14}$$

where ΔT is temperature difference between the two heights of measurement (z_1 and z_2); Federer (1970) presented Eq. 8.15.

$$k_h = \frac{k u_* z}{\Phi_h} \tag{8.15}$$

In this chapter, Lake Okeechobee in south Florida is selected to show the application of different evaporation estimation methods as meteorological data is available measured in the lake on platforms. Figure 8.5 depicts south Florida and Lake Okeechobee with weather monitoring sites. Site L006 was used in this analysis.

The platform with monitoring stations and close-up of the sensors is shown in Fig. 8.6a, b. Monitoring parameters in Lake Okeechobee at weather station L006 are shown in Table 8.2.

To demonstrate the application of the energy balance method, net solar radiation is borrowed from a land-based weather station in the vicinity of the lake (ENR308).

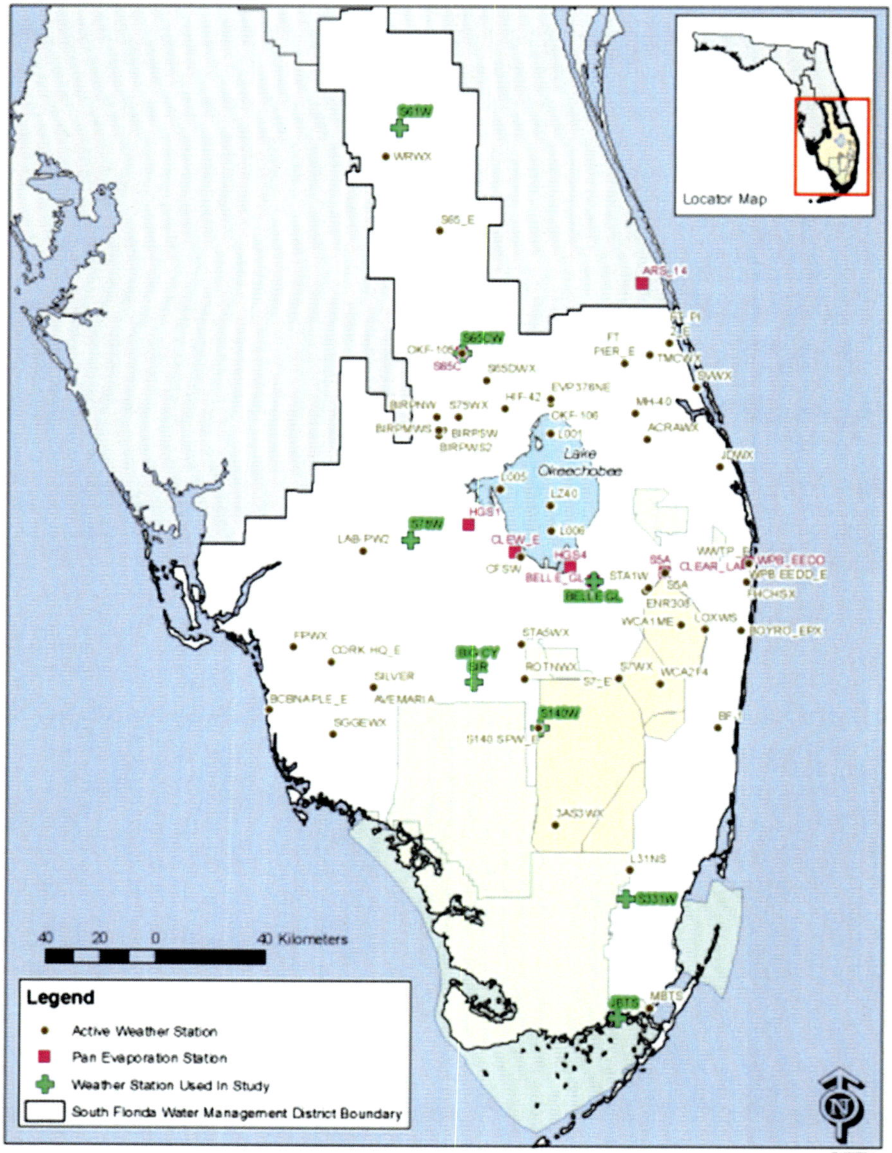

Fig. 8.5 South Florida and Lake Okeechobee with weather-monitoring sites (Abtew et al. 2011)

Equations 8.11 and 8.13 were applied for estimating the heat transfer coefficient using the meteorological parameter observations in Lake Okeechobee. Sensible heat is computed using Eq. 8.8. Two parameters to be estimated are the friction velocity (u_*) and the inverse of the wind speed gradient (dz/du). In order to estimate u_*, many more parameters have to be estimated. Table 8.3 depicts equations used to estimate parameters required to derive u_*, k_h, and H.

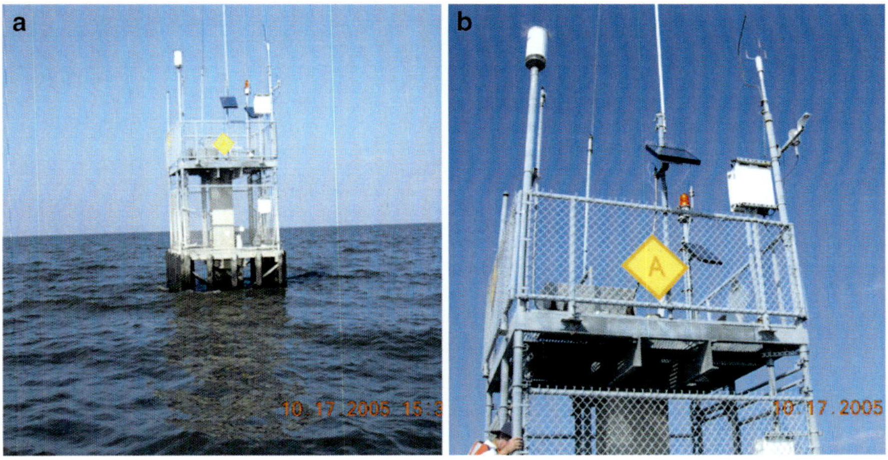

Fig. 8.6 (**a**) Weather station tower in Lake Okeechobee (Photograph provided by South Florida Water Management District), (**b**) Close-up of weather station tower sensors in Lake Okeechobee (Photograph provided by South Florida Water Management District)

Table 8.2 Monitoring parameters at station L006 in Lake Okeechobee (average water surface elevation is 4.7 m NGVD29)

Parameter	Unit	Height above water level	Frequency
Water temperature	°C	15 cm below water level	15 min
Water temperature	°C	1.5 m above lake bottom	15 min
Water temperature	°C	30 cm above lake bottom	15 min
Air temperature	°C	At 9.8 m − lake elevation	15 min
Humidity	%	At 9.8 m − lake elevation	15 min
Wind speed	mph	At 11.7 m − lake elevation	15 min
Wind direction	Degrees	At 11.7 m − lake elevation	15 min
Wind gust (maximum)	mph	At 11.7 m − lake elevation	10 s
Atmospheric pressure	kPa	At 9.8 m − lake elevation	15 min
Solar radiation	kw m^{-2}	At 9.8 m − lake elevation	15 min
Rain	Inches	At 11.7 m − lake elevation	15 min

The energy balance method was applied for 1 day, May 1, 1998, with 15 min of meteorological data measured inside the lake except net solar radiation (Table 8.2). Water heat flux (G in kJ m^{-2}) is computed by Eq. 8.16. Results of the analysis are presented in Table 8.4 as an hourly average from 6:00 a.m. to 7:00 p.m.:

$$G = c_s d_w (T_n - T_{n-1})^* \frac{1,000}{86,400} \tag{8.16}$$

where c_s is water heat capacity (4.18 MJ m^{-3} °C^{-1}) and d_w is water depth where the top 30-cm water depth was used for change in storage computation with water

Table 8.3 Supplementary parameters for energy balance evaporation estimation

Parameter	Equation	Remarks
u_*	$u_* = \frac{uk}{\ln((z-d)/z_o)}$	u = wind speed (m s^{-1}) at 2-m height k = von Karman constant (0.41) z = height of wind speed measurement (m) d = displacement height (m) z_o = roughness height (m)
d (Abtew et al. 1989)	$d = 0.5$ h	h = average wave height (m)
z_o (Abtew et al. 1989)	$Z_o = 0.13\ (h–d)$	–
h (Linsley and Franzini 1979)	$h = 0.005\ u^{1.06}\ F^{0.47}$	u = wind speed (km h^{-1}) at 6.8-m height F = fetch (km)
dz	Average of two heights of winds speed measurement	In m
du	Average of change in wind speed from water surface to 2 m and from 2 to 6.8 m	In m s^{-1}
dT	Change in temperature between water temperature at 15-cm depth and air temperature at 4.7 m	°C
θ (Federer 1970)	$\theta_* = \frac{\Delta T k}{\ln(z_2/z_1)}$	$z_2 = 5.2$ m and $z_1 = 0.5$ m Height raised by 0.5 m to match k_h computed by Eq. 8.11 and to avoid dividing by zero or small height at the surface

temperature measured at 15-cm depth in the lake. T_n and T_{n-1} are water temperature on day $n-1$ and n. Advection energy is assumed negligible with the assumption that inflow and outflow temperature is the same as lake surface water temperature.

Based on the energy balance, lake evaporation for 24 h on May 1, 1998, was 4.04 mm. Energy balance components are shown for May 1, 1998, from 6:00 a.m. to 7:00 p.m. (Fig. 8.7a). The 15-min lake evaporation in mm is shown in Fig. 8.7b. A limitation of the energy balance method is that the ability of the environment to transfer or hold water vapor is not accounted. These limitations are shown by the vapor pressure deficit pattern over Lake Okeechobee from the 1-day (May 1, 1998) energy balance analysis (Fig. 8.7c).

8.2.4 Mass Transfer Method

Mass transfer models are based on estimating the net transport of water vapor from the lake surface to the atmosphere (Ikebuchi et al. 1988). The methods are based on Dalton's law where vapor transfer from an evaporating surface is proportional to

Table 8.4 Hourly average energy flux and other parameters on May 1, 1998

Time	u @ 6.61 mm s^{-1}	u @ 2 mm s^{-1}	Wave height m	ΔT °C	u_* m s^{-1}	k_h (Eq. 8.11) m^2 s^{-1}	R_n kJ m^{-2}	G kJ m^{-2}	H kJ m^{-2}	E kJ m^{-2}
6:00	6.93	4.91	0.856	0.895	0.603	0.248	0.004	0	0.058	−0.054
7:00	6.15	4.45	0.755	1.035	0.522	0.207	0.154	0	0.056	0.098
8:00	5.48	4.02	0.668	0.518	0.452	0.16	0.255	−0.002	0.024	0.232
9:00	3.48	2.67	0.412	0.089	0.26	0.09	0.441	0	0.002	0.439
10:00	4.82	3.58	0.583	−1.34	0.386	0.144	0.354	0	−0.051	0.404
11:00	4.11	3.11	0.493	−1.1	0.32	0.115	0.432	0.006	−0.034	0.461
12:00	2.05	1.62	0.236	−0.84	0.14	0.044	0.621	0.005	−0.008	0.624
13:00	1.42	1.15	0.16	−2.38	0.091	0.026	0.54	−0.009	−0.015	0.563
14:00	0.43	0.36	0.045	−4.685	0.023	0.006	0.413	−0.001	−0.007	0.421
15:00	2.97	2.26	0.352	−3.601	0.225	0.08	0.19	0.001	−0.075	0.264
16:00	7.37	5.15	0.916	−1.608	0.655	0.278	0.079	0.004	0.113	0.187
17:00	9	6.06	1.13	−0.825	0.838	0.373	0.013	−0.001	−0.077	0.09
18:00	11.68	7.26	1.49	1.143	1.169	0.576	−0.02	−0.001	0.163	0.252

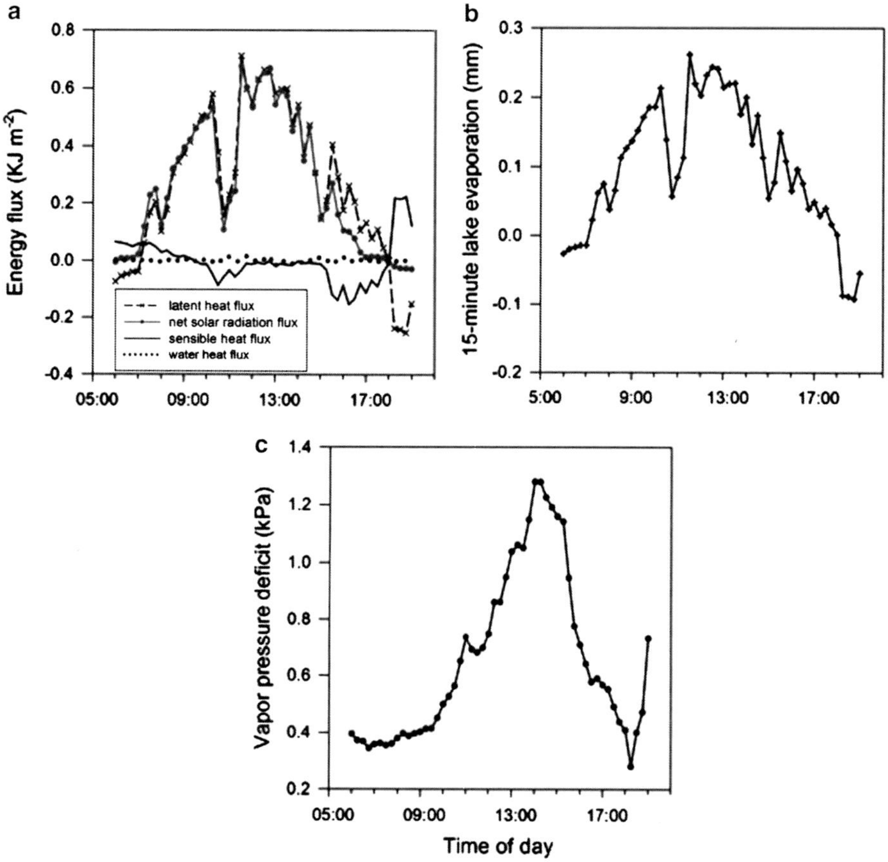

Fig. 8.7 (**a**) Energy balance components, (**b**) 15-min lake evaporation, and (**c**) vapor pressure deficit (May 1, 1998)

wind velocity and vapor pressure deficit over the surface. By combining the sheer stress and latent heat flux, the mass and momentum transfer equations produce a mass transfer equation (Eq. 8.17) for estimating evaporation (E) (Singh 1989):

$$E = \rho u_*^2 \frac{k_w(q_2 - q_1)}{k_m(u_2 - u_1)} \tag{8.17}$$

where $(q_2 - q_1)$ is difference in specific humidity (dimensionless) at heights z_2 and z_1 above the water surface and $(u_2 - u_1)$ is wind speed difference between the heights z_2 and z_1. A mass transfer model that was originally developed by Harbeck (1962) was applied by Hostetler and Bartlein (1990) to estimate evaporation for Harney–Malheur Lake in Oregon (Eq. 8.18).

$$E = N_o u_2 (e_o - e_a) \tag{8.18}$$

where E is in mm, N_o is an empirically determined mass transfer coefficient (mm s m^{-1} kPa^{-1}), u_2 is wind speed at 2-m height above the lake surface, e_o is saturation vapor pressure at the lake surface (kPa), and e_a is ambient vapor pressure of the air (kPa). The mass transfer coefficient N_o is computed for large lakes from lake surface area, A (km^2), by Eq. 8.19 (Shuttleworth 1993).

$$N_o = 2.909\ A^{-0.05} \tag{8.19}$$

This method was applied for the same day the energy balance method was applied for Lake Okeechobee (May 1, 1998). The lake surface area is 1,732 km^2. The computed mass transfer coefficient N_o is 2.00, and the average 2-m height wind speed above the lake surface was 4.18 m s^{-1}. Evaluating the performance of the mass transfer method for evaporation estimation in a semiarid region of India, the coefficient, N_o, was found to be 2.35 (Ali et al. 2007). The average vapor pressure difference calculated as described above is 0.453 kPa. Using the mass transfer method (Eq. 8.18), the estimated evaporation for Lake Okeechobee on May 1, 1998, is 3.79 mm. The limitation with the mass transfer method is that energy required for evaporation is not considered. Increase in wind speed and vapor pressure deficit result in extremely high evaporation. Both the energy balance and mass transfer methods were applied for the full month of May 1998 using daily average meteorological data. The monthly mean lake evaporation by the energy balance and mass transfer methods was 4.77 and 4.86 mm day^{-1}, respectively. When daily evaporation estimates are compared, the limitations of both methods are clearly shown. Figure 8.8a depicts wind speed at 2-m height and vapor pressure deficit over Lake Okeechobee for the month of May 1998. Figure 8.8b depicts daily lake evaporation estimates by the energy balance and the mass transfer methods.

8.2.5 *The Penman Method*

Penman in 1948 derived a combination equation to estimate evaporation. The method combines the energy required to cause evaporation and the mechanisms required to remove vapor from the evaporating surface (Jensen et al. 1990). Vapor pressure deficit combined with wind speed creates the condition for vapor movement from higher saturation to lower saturation zones and the resistance of vapor movement from water to air. The energy balance between net radiation, sensible heat flux, and change in heat storage results in energy available for evaporation. The Penman combination equation (Eq. 8.20) that is applied to compute potential evapotranspiration or open water evaporation is given as follows (Shuttleworth 1993; Valiantzas 2006):

$$\mathrm{ET} = \frac{1}{\lambda}\frac{\Delta(R_n - G) + \gamma 6.43(f(u))(e_a - e_d)}{(\Delta + \gamma)} \tag{8.20}$$

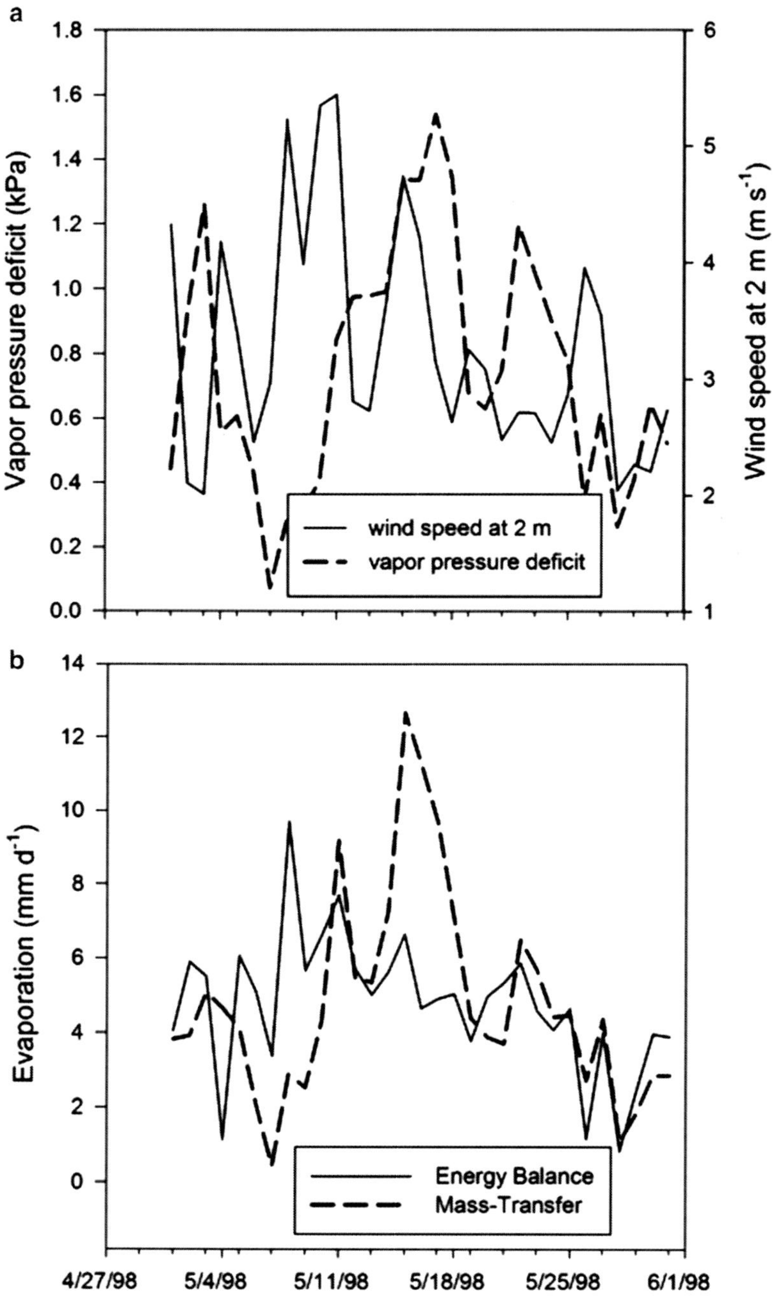

Fig. 8.8 (**a**) Wind speed at 2-m height and vapor pressure deficit, (**b**) daily evaporation estimates by energy balance and mass transfer methods

where E is evaporation in mm day^{-1}, R_n is net radiation (MJ m^2 day^{-1}), G is water heat flux (MJ m^2 day^{-1}), is slope of vapor pressure curve (kPa °C^{-1}), γ is psychrometric constant (kPa °C^{-1}), e_s is saturation vapor pressure, e_d is actual vapor pressure, $(e_s - e_d)$ is vapor pressure deficit computed using air temperature, and $f(u)$ is wind function expressed by Eq. 8.21:

$$f(u) = a_w + b_w u_2 \tag{8.21}$$

where a_w and b_w are wind function coefficients and u_2 is wind speed at 2-m height (m s^{-1}). In the original Penman equation, $a_w = 1$ and $b_w = 0.536$ (Valiantzas 2006). Realizing that these coefficients are site dependent, J.W. Wright, USDA, Kimberly, Idaho, developed equations to estimate wind coefficients based on a normal probability density function to reflect seasonal variation with the general form shown by Eqs. 8.22 and 8.23 for northern latitudes (Allen et al. 1989; Shuttleworth 1993):

$$a_w = c_1 + c_2 \exp\left\{-\left[\frac{J - 173}{58}\right]^2\right\} \tag{8.22}$$

$$b_w = c_3 + c_4 \exp\left\{-\left[\frac{J - 243}{80}\right]^2\right\} \tag{8.23}$$

where c_1, c_2, c_3, and c_4 are site-dependent coefficients. For Kimberly, Idaho, c_1, c_2, c_3, and c_4 are 0.4, 1.4, 0.605, and 0.345, respectively. Coefficients of 0.1, 3.0, 0.04, and 0.2 were developed for south Florida based on least square fit of lysimeter measured evaporation and model estimated data (Abtew 1996). The Penman method was also applied to estimate lake evaporation for May 1998. The average daily evaporation was 4.47 mm day^{-1}.

8.2.6 The Simple Abtew Method

Comparison of lysimeter measurements of open water evaporation and wetland vegetation evapotranspiration showed that there is no significant difference between evaporation from shallow lakes, evapotranspiration from wetland vegetation, and potential evapotranspiration in south Florida. It was also shown that most of the variance (73%) in evaporation and evapotranspiration in south Florida is explained by variation in solar radiation. A simple equation was calibrated to estimate daily lake evaporation, wetland evapotranspiration, or potential evapotranspiration in south Florida (Abtew 1996). Equation 8.24 is also cited as the Abtew equation and simple Abtew equation in published literature (Abtew 1996; Xu and Singh 2000; Abtew et al. 2003; Delclaux and Coudrain 2005; Oudin et al. 2005; Shoemaker and Sumner 2006; Melesse et al. 2009; Zhai et al. 2009; Enku et al. 2011; Setegn et al. 2011). The advantage of the simple equation is that it uses only a single input parameter, R_s:

$$\mathrm{ET} = K_1 \frac{R_s}{\lambda} \tag{8.24}$$

where ET is daily evapotranspiration from wetland or shallow open water or potential evapotranspiration (mm day^{-1}), R_s is solar radiation (MJ m^{-2} day^{-1}), λ is latent heat of vaporization (MJ kg^{-1}), and K_1 is a dimensionless coefficient (0.53). The mm day^{-1} unit is derived from the fact that a kilogram of water is 1,000 cc (10^6 mm^3) and a square meter is 10^6 mm^2. The simple Abtew method was also applied to estimate lake evaporation for May 1998. The average daily lake evaporation estimate was 4.68 mm day^{-1}. The simple Abtew method is currently applied to acquire daily evaporation data for Lake Okeechobee in south Florida and stored in a widely accessed hydrometeorological database, DBHYDRO. It was also successfully applied to estimate evaporation from Lake Ziway in the Ethiopian Rift Valley (Melesse et al. 2009). Compared to eight evaporation models, the Simple Method provided the best result in estimating evaporation from Lake Titicaca (Delclaux and Coudrain 2005). Lake Titicaca is located in the Southern Hemisphere, between Bolivia and Peru at 3,810-m elevation (16°S 69°W).

8.2.7 Solar Radiation–Maximum Temperature Method

In order to include air temperature in evaporation estimation, an equation was developed and calibrated using lysimeter measurements (Abtew 1996). Maximum air temperature in °C (T_{max}) is added to Eq. 8.24 with a calibration coefficient, K_3 (°C), Eq. 8.25. This method was also applied to estimate lake evaporation for May 1998. The average daily lake evaporation estimate was 4.63 mm day^{-1}:

$$\mathrm{ET} = \frac{1}{k_3} \frac{R_s}{\lambda} T_{max} \tag{8.25}$$

8.2.8 Modified Turc Equation

The original Turc equation for potential evapotranspiration estimation for humid regions is similar to Eq. 8.26 but uses daily average temperature. The Turc equation was modified by using daily maximum air temperature (°C) as this gave better fit to measured lysimeter data in south Florida, a humid and warm subtropical region (Abtew 1996, 2001). The coefficient K_2 is similar to the original Turc value of 0.013, which was originally recommended for humid region:

$$\mathrm{ET_P} = K_2 \frac{(23.89R_s + 50)T_{\max}}{(T_{\max} + 15)} \tag{8.26}$$

where E is evaporation in mm and R_s is solar radiation in MJ m^{-2} day^{-1}. K_2 has unit mm MJ^{-1} m^2 day in order to balance the units. The modified Turc equation was also applied to estimate lake evaporation for May 1998. The average daily lake evaporation estimate was 4.87 mm day^{-1}.

8.2.9 Priestley–Taylor Method

The Priestley–Taylor equation is a simplified form of the Penman equation, where the aerodynamic component is left out of the equation. A coefficient (α), with a value of greater than 1 (1.26), is included to make up for the loss of the aerodynamic component of evaporation. This method is widely used to estimate evaporation sometimes with modified coefficient (α) (Reis and Dias 1998; Abtew 1996). The Priestley–Taylor equation is presented by Eq. 8.27:

$$\mathrm{ET} = \frac{\alpha}{\lambda} \frac{\Delta R_n}{(\Delta + \gamma)} (R_n - G) \tag{8.27}$$

The Priestley–Taylor equation was also applied to estimate lake evaporation for May 1998 with α value of 1.26. The average daily lake evaporation estimation was 4.26 mm day^{-1}. The average daily lake evaporations estimated by the Penman, simple Abtew, solar radiation–maximum temperature, modified Turc, and Priestley–Taylor methods are 4.47, 4.68, 4.63, 4.87, and 4.26 mm day^{-1}, respectively. Figure 8.9 depicts daily evapotranspiration measurement by the Penman, simple Abtew, solar radiation–maximum temperature, modified Turc, and Priestley–Taylor methods.

8.2.10 Energy Balance–Bowen Ratio Method (EBBR)

Estimation of sensible heat, H, in the energy balance method is challenging as shown in Sect. 8.2.3. As shown in Eq. 8.8, temperature change with height and a transfer coefficient is required to estimate H. The EBBR method substitutes the Bowen ratio in the energy equation in place of H (Eq. 8.28). The Bowen ratio (β) is the ratio of sensible heat to latent heat flux (Eq. 8.29):

$$\lambda E = \frac{R_n - G}{1 + \beta} \tag{8.28}$$

$$\beta = \frac{H}{\lambda E} = \gamma \frac{\Delta T}{\Delta e} \tag{8.29}$$

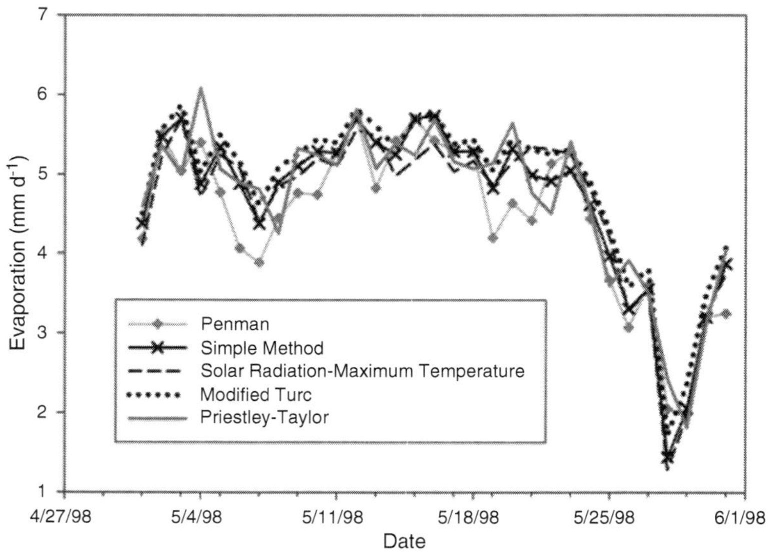

Fig. 8.9 Daily lake evaporation estimation by the Penman, simple Abtew, solar radiation–maximum temperature, modified Turc, and Priestley–Taylor methods

The Bowen ratio estimation requires temperature and vapor pressure measurements at two heights over the water surface. Different approaches have been presented to avoid measurements at two heights: replacing with water and air temperature measurements and associated saturation vapor, actual vapor pressure, and air pressure. Referring to studies at Lake Mead and Lake Eucumbene, Omar and El-Bakry (1981) applied a different format (Eq. 8.30) in their estimation of evaporation from Lake Nasser, Aswan Dam. Stannard and Rosenberry (1991) credited the Bowen ratio equation to E.R. Anderson and Lake Hefner, Oklahoma, evaporation study. Both formats use a constant and air pressure in place of γ. The Bowen ratio estimation equation (Eq. 8.30) with analysis is presented by Reis and Dias (1998):

$$\beta = \gamma \frac{(T_s - T_a)}{(e_s - e_d)} \tag{8.30}$$

where T_s is lake surface water temperature (°C), T_a is air temperature over the lake (°C), e_s is saturation vapor pressure corresponding to T_s (kPa), and e_d is the air actual vapor pressure corresponding to T_a (kPA). Lake Okeechobee temperature, pressure, and vapor pressure data was used to compute daily evaporation for May 1998. The average daily evaporation with the EBBR method (Eqs. 8.28 and 8.30) is 4.47 mm. Daily evaporation estimates are shown in Fig. 8.10 comparing EBBR, Penman, and the simple Abtew methods. Except at the beginning of the month, the EBBR method has given comparable estimates to the Penman, simple Abtew, and other methods.

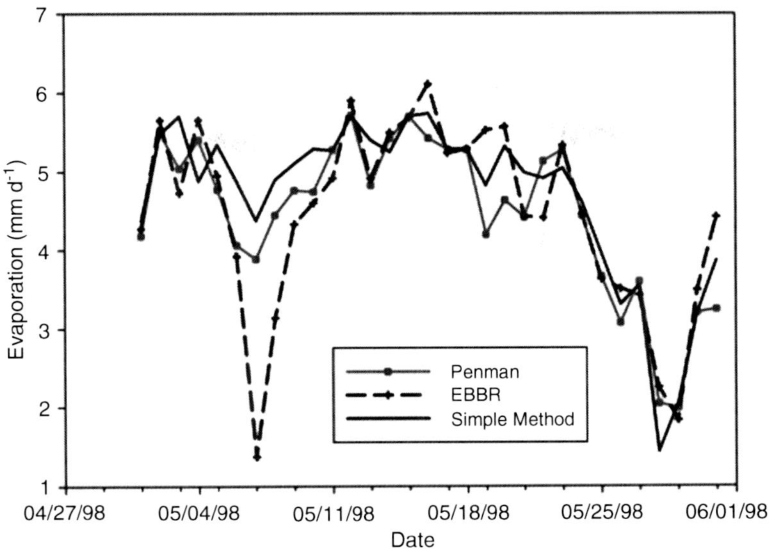

Fig. 8.10 Daily lake evaporation estimation by the Penman, EBBR, and simple Abtew methods

The EBBR method fails to produce reasonable estimates of evaporation when there is a sudden drop in air temperature compared to water temperature or vice versa.

8.3 Summary

Lake evaporation estimation or measurement has been challenging and various methods have been applied for different lakes. Method selection should be dependent on location or environment of the lake and available input data. For lakes located in arid areas, advection energy needs to be accounted for. For tropical and subtropical lakes, the most dominant parameter, solar radiation, may be enough to estimate evaporation. Adding temperature average or maximum into the equation could improve the estimate. The mass transfer method has the potential to produce out of range estimates due to high influence of wind speed on the equation. Relatively, energy balance methods produce within range values, as the available energy limits the maximum evaporation that could occur. The energy balance–Bowen ratio (EBBR) method is susceptible to error of estimation when daily difference in lake surface water and air temperatures is large. The simple Abtew method and the solar radiation–maximum temperature methods use minimum input and produce competitive results for south Florida and other locations where applied. In tropical and subtropical areas, such methods can be tested and successfully applied. Methods that do not use net radiation bypass the technical challenges of acquiring good quality net radiation data for longer time periods. To compare the

Table 8.5 Mean daily evaporation, standard deviation, and range for Lake Okeechobee for the month of May 1998

Method	Mean (mm)	Stdev (mm)	Range
Energy balance	4.77	1	0.83–9.69
Mass transfer	4.86	2.81	0.4–12.68
Penman	4.47	1	1.99–5.71
Simple Abtew method	4.68	1.05	1.44–5.74
Solar radiation–maximum temperature method	4.63	1.04	1.28–5.70
Modified Turc equation	4.87	0.99	1.73–5.87
Priestley–Taylor equation	4.26	0.88	1.65–5.41
Energy balance–Bowen ratio method	4.47	1.19	1.37–6.11

difference between evapotranspiration methods, good quality shorter period data should be preferred than lower quality but longer period data. Otherwise, error in data will bias the results. Table 8.5 summarizes evaporation estimation for Lake Okeechobee in south Florida for the month of May 1998. The energy balance and mass transfer methods have wider ranges and out of range values reflecting the inherent deficiencies of the two methods.

References

Abtew W (1996) Evapotranspiration measurements and modeling for three wetland systems in South Florida. J Am Water Resour Assoc 127(3):140–147

Abtew W (2001) Evaporation estimation for Lake Okeechobee in South Florida. J Irrig Drain Eng 127(3):140–147

Abtew W, Gregory JM, Borrelli J (1989) Wind profile: estimation of displacement height and aerodynamic roughness. Trans ASAE 32(2):521–527

Abtew W, Obeysekera J, Ortiz MI, Lyons D, Reardon A (2003) Evapotranspiration estimation for South Florida. In: Bizier P, DeBarry P (eds) Proceedings of the world water and environmental congress 2003. ASCE, Reston

Abtew W, Iricanin N, Obeysekera J (2011) Pan evaporation and potential evapotranspiration trends in South Florida. Hydrol Process 25:958–969

Ali S, Ghosh NC, Singh R (2007) Evaluating best evaporation estimate model for water surface evaporation in semi-arid region, India. Hydrol Process 22(8):1093–1106

Allen RG, Jensen ME, Wright JL, Burman RD (1989) Operational estimates of reference evapotranspiration. Agron J 81:650–662

Allen RG, Walter IA, Elliott RL, Howell TA, Itenfisu D, Jensen ME, Snyder RL (2005) The ASCE standard reference evapotranspiration equation. ASCE, Reston

Boyd C (1985) Pond evaporation. Trans Am Fish Soc 114:299–303

Delclaux F, Coudrain A (2005) Optimal evaporation models for simulation of large lake levels: application to Lake Titicaca, South America. Geophys Res Abstr 7:53–65

Enku T, van der Tol C, Gieske ASM, Rientjes THM (2011) Evapotranspiration modeling using remote sensing and empirical models in the Fogera floodplain, Ethiopia. In: Melesse A (ed) Nile River Basin: hydrology, climate and water use. Springer, Dordrecht, p 163, 170

FAO (1977) Crop water requirements. FAO, Rome

Federer CA (1970) Measuring forest evapotranspiration-theory and practice. USDA Forest Service Research Paper NE-165. US Department of Agriculture, Washington, DC

Gifford RM, Roderick M, Farquhar GD (2007) Evaporative demand: does it increase with global warming? Glob Change Newsl 69:21–23

Harbeck GE Jr (1962) A practical field technique for measuring reservoir evaporation utilizing mass-transfer theory. Geological Survey Professional Paper 272-E. US Government Printing Office, Washington, DC

Hostetler SW, Bartlein PJ (1990) Simulation of lake evaporation with application with to modelling lake level variation of Harney-Malheur Lake, Oregon. Water Resour Res 26(10):2603–2612

Ikebuchi S, Seki M, Ohtoh A (1988) Evaporation from Lake Biwa. J Hydrol 102:427–449

Jacovides C, Kerkides P, Papiaoannou G, Smith FB (1992) Evaluation of the profile and the resistance method for estimation of surface fluxes of momentum, sensible and latent heat. Theor Appl Climatol 45:145–154

Jensen ME (1974) Consumptive use of water and irrigation water requirements. ASCE, New York

Jensen ME, Burman RD, Allen RG (eds) (1990) Evapotranspiration and irrigation water requirements. ASCE manuals and reports on engineering practice No. 70. ASCE, New York

Jin K, James RT, Ling W, Louks DP, Park RA (1998) Assessing Lake Okeechobee eutrophication with water-quality models. J Water Resour Plan Manage 124(1):22–30

Katul GG, Parlange MB (1992) A Penman-Brutsaert model for surface evaporation. Water Resour Res 28(1):121–126

Kohler MA, Nordenson T, Baker D (1959) Evaporation maps for the United States. US Weather Bureau Technical Paper 37, Washington, DC

Langbein WB (1951) Research on evaporation from lakes and reservoirs. IASH-AISH Publ 34, pp 409–425

Lenters JD, Kratz TK, Bowser CJ (2005) Effects of climate variability on lake evaporation: results from a long-term energy budget study of Sparkling Lake, northern Wisconsin (USA). J Hydrol 208:168–195

Linacre ET (1993) Data-sparse estimation of lake evaporation, using a simplified Penman equation. Agric Forest Meteorol 64:237–256

Linsley RK, Franzini JB (1979) Water-resources engineering. McGraw-Hill, New York

Melesse A, Abtew W, Dessalegne T (2009) Evaporation estimation of Rift Valley Lakes: comparison of models. Sensor J 9:9603–9615. doi:10.3390/s91209603

Monteith JL (1973) Principles of environmental physics. Edward Arnold, London

Morton FI (1986) Practical estimates of lake evaporation. J Clim Appl Meteorol 25:371–387

Omar MH, El-Bakry MM (1981) Estimation of evaporation from the lake of the Aswan High Dam (Lake Nasser) based on measurements over the lake. Agric Meteorol 23:293–308

Oudin L, Hervieu F, Michel C, Perrin C, Andreassian V, Anctil F, Loumagne C (2005) Which potential evapotranspiration input for a lumped model part 2-towards a simple and efficient potential evapotranspiration model for rainfall-runoff modeling. J Hydrol 303:290–306

Reis RJD, Dias NL (1998) Multi-season lake evaporation: energy-budget estimates and CRLE model assessment with limited meteorological observations. J Hydrol 208:135–147

Roberts WJ, Stall JB (1967) Lake evaporation in Illinois. Illinois State Water Survey, Urbana

Setegn SG, Chowdary VM, Mal BC, Yohannes F, Kono Y (2011) Water balance study and irrigation strategies for sustainable management of a tropical Ethiopian lake: a case study of Lake Alemaya. Water Resour Manage 25(9):2081–2107. doi:10.1007/s11269-011-9797-y

Shoemaker WB, Sumner DM (2006) Alternate corrections for estimating actual wetland evapotranspiration from potential evapotranspiration. Wetlands 26(2):528–543

Shuttleworth WJ (1993) Chapter 4: Evaporation. In: Maidment DR (ed) Handbook of hydrology. McGraw-Hill, Inc., New York

Singh VP (1989) Hydrologic systems-watershed modelling, vol II. Prentice-Hall, Engelewood Cliffs

Stannard DI (1993) Comparison of Penman-Monteith, Shuttleworth-Wallace, and modified Priestly-Taylor evapotranspiration models for wildland vegetation in semiarid rangeland. Water Resour Res 29(5):1379–1392

Stannard DI, Rosenberry DO (1991) A comparison of short-term measurements of lake evaporation using eddy correlation and energy budget methods. J Hydrol 122:15–20

Valiantzas JD (2006) Simplified versions for the Penman evaporation equation using routine weather data. J Hydrol 331:690–702

Webb EK (1966) A pan-lake evaporation relationship. J Hydrol 4:1–11

Winter TC, Rosenberry DO (1995) Evaluation of 11 equations for determining evaporation from a small lake in the north central United States. Water Resour Res 31(4):983–993

Xu CY, Singh VP (2000) Evaluation and generalization of radiation-based methods for calculating evaporation. Hydrol Process 14:339–349

Zhai L, Feng Q, Li Q, Xu C (2009) Comparison and modification of equations for calculating evapotranspiration (ET) with data from Gansu Province, Northwest China. Irrig Drain 58:1–14. doi:10.1002/ird.502

Chapter 9
Reference and Crop Evapotranspiration

Abstract Advancements have been made in estimating potential and reference evapotranspiration. Actual crop evapotranspiration estimation still is a challenge due to limited availability of local crop coefficients and large-scale variation in field conditions. In this chapter, application and comparison of currently used potential and reference evapotranspiration estimation methods are presented. Crop coefficient availability is discussed and a general seasonal pattern of crop coefficients is presented.

Keywords Reference evapotranspiration • Crop evapotranspiration • Potential evapotranspiration • Crop coefficient • Reference evapotranspiration estimation methods • Canopy resistance • Aerodynamic resistance

9.1 Introduction

Most of the chapters in this book address evaporation from open water surface and evapotranspiration (ET) from wetland surfaces where water is not a limiting factor. Evapotranspiration estimation from crop surfaces or other vegetation where water is a limiting factor is far more challenging. The approach adopted is the estimation of reference evapotranspiration from a hypothetical well-watered vegetated surface of known height (Allen et al. 2005; Smith 1991) and deriving crop evapotranspiration using crop coefficients. The reference well-watered crop is usually referenced as well-watered alfalfa of 12-cm height. At times, reference evapotranspiration is interchangeably used as potential evapotranspiration. But they are not similar. Reference evapotranspiration equations are parameterized to generate reference ET estimates without regard to the maximum limits to evapotranspiration at the location. Meaningful actual ET rates are derived by applying one or more coefficients. Potential ET is the maximum ET that could occur at a site under the prevailing meteorological conditions. Actual crop evaporation estimates are derived from reference ET through application of seasonally varying crop coefficients.

W. Abtew and A. Melesse, *Evaporation and Evapotranspiration: Measurements and Estimations*, DOI 10.1007/978-94-007-4737-1_9,

Pan evaporation has also been used as reference ET, and crop ET is derived applying two coefficients: potential ET coefficient (K_p) and crop coefficient (K_c) (Abtew and Sculley 1991).

9.2 Reference Evapotranspiration

FAO-24 (Doorenbos and Pruitt 1977) recommended the Blaney–Criddle method for reference evapotranspiration estimation. This method is presented in Chap. 6 with application to a region. Later, FAO (Smith 1991) recommended the use of the Penman–Monteith method for reference ET estimation. It provided estimation equations for r_c (crop canopy resistance) and r_a (aerodynamic resistance) and other derived and estimated parameters. The Penman–Monteith method is a combination method where part of the ET is due to radiation terms, ET_{rad}, and part is due aerodynamic terms, ET_{aero} (Eq. 9.1):

$$ET = ET_{rad} + ET_{aero} \tag{9.1}$$

The Penman–Monteith equation is given as follows (Eq. 9.2):

$$ET = \frac{\Delta(R_n - G) + \rho c_p (e_s - e_d)\frac{1}{r_a}}{\Delta + \gamma\left(1 + \frac{r_c}{r_a}\right)} \tag{9.2}$$

where ET is latent heat flux of evaporation (kJ m^{-2} s^{-1}), R_n is net radiation flux (kJ m^{-2} s^{-1}), is the slope of the vapor pressure curve (kPa °C^{-1}), γ is psychrometric constant (kPa °C^{-1}), G is soil heat flux (kJ m^{-2} s^{-1}), ρ is atmospheric density (kg m^{-3}), c_p is specific heat of moist air (kJ kg^{-1} °C^{-1}), $(e_a - e_d)$ is vapor pressure deficit (kPa), r_c is canopy resistance, and r_a is aerodynamic resistance. This method has the most measured, derived, and estimated inputs as shown in Table 6.3 in Chap. 6. Resistance factors are computed as follows.

9.2.1 Crop Canopy Resistance (r_c)

Crop canopy resistance (r_c) is computed from average daily (24 h) stomatal resistance (r_l) of a single leaf estimated as 100 s m^{-1} and leaf area index (LAI) as shown in Eq. 9.3 (Allen et al. 1989):

$$r_c = \frac{r_l}{0.5\text{LAI}} \tag{9.3}$$

Leaf area index is estimated from reference crop height (h_c). Equation 9.4 is for clipped grass of 12-cm height, and Eq. 9.5 is for alfalfa and other field crops with height ranging from 10 to 50 cm:

$$\mathrm{LAI} = 24\, h_c \tag{9.4}$$

$$\mathrm{LAI} = 5.5 + 1.5\, \ln(h_c) \tag{9.5}$$

For grass of 12-cm height, LAI is 2.88 and r_c is 70 s m^{-1}. Due to inconsistency in reference ET estimation from differences in input data quality and parameter estimation, a more standardized method was needed for practical applications.

9.2.2 *Aerodynamic Resistance (r_a)*

The aerodynamic resistance has been commonly presented as mainly a function of surface characteristics and wind speed. Equation 9.6 (Allen et al. 1989) has been in use:

$$r_a = \frac{\ln \frac{(z-d)}{z_{om}}}{k^2} \times \frac{\ln \frac{(z_h-d)}{z_{oh}}}{u_z} \tag{9.6}$$

where r_a is aerodynamic resistance (s m^{-1}), z is the height of wind measurement (m), z_h is the height of air temperature and humidity measurement (m), d is displacement height (m), z_{om} is roughness length for momentum transfer, z_{oh} is roughness length for vapor and heat transfer, k is the von Karman constant for turbulent diffusion (0.41), and u_z is wind speed measurement at height z in m s^{-1}. Application of Eq. 9.6 on daily basis over a year in south Florida resulted in a mean r_a of 83.3 s m^{-1}.

There are several equations to estimate displacement height (d) and aerodynamic roughness (z_o) as shown in Abtew et al. (1989). Applying the author's methods (Eqs. 9.7 and 9.8), wind speed is estimated at 2 m (Abtew et al. 1989). z_{oh} is estimated by Eq. 9.9 (Allen et al. 1989):

$$d = F_c h_c \tag{9.7}$$

where F_c is fraction of surface cover and h_c is average height of cover.

$$z_{om} = 0.13(h_c - d) \tag{9.8}$$

$$z_{oh} = 0.1 z_{om} \tag{9.9}$$

For the reference crop of 0.12-m grass, d is 0.08 cm, z_{om} is 0.015 m, and z_{oh} is 0.0015 m. For temperature and humidity measurements at 2-m height, aerodynamic resistance (r_a) is estimated by Eq. 9.10 after replacing the values for each variable:

$$r_a = \frac{200}{u_2} \tag{9.10}$$

where u_2 is wind speed measurement at 2-m height (m s^{-1}). Computation of daily variable r_a is shown in Chap. 6. Once average values are determined for the resistance factors, the consistent application of the Penman–Monteith equation rests on the quality of net radiation, air temperature, and humidity data.

9.3 The ASCE Standardized Reference Evapotranspiration Equation

The ASCE standardized reference evapotranspiration equation was formulated for the purpose of standardizing reference ET estimation and improving transferability of crop coefficients (Allen et al. 2005). An additional objective was to simplify and clarify the presentation and application of the reference ET estimation method. The equation was developed to calculate reference ET for short crop (0.12 m similar to clipped cool-season grass) and tall crop (0.5 m similar to full-cover alfalfa). Different coefficients and variables are used for each surface. The basic equation is given by Eq. 9.11:

$$\mathrm{ET_{sz}} = 0.408 \frac{\Delta(R_n - G) + \gamma \frac{C_n}{T+273} u_2 (e_s - e_d)}{\Delta + \gamma(1 + C_d u_2)} \tag{9.11}$$

where $\mathrm{ET_{sz}}$ (mm day^{-1}) is standardized reference crop evapotranspiration for short or tall crop, R_n is net radiation (MJ m^{-2} day^{-1}) , Δ is the slope of the vapor pressure curve (kPa °C^{-1}), γ is psychrometric constant (kPa °C^{-1}), G is soil heat flux density (MJ m^{-2} day^{-1}), T is mean daily air temperature at 1.5–2.5 m, u_2 is mean daily wind speed at 2-m height (m s^{-1}), e_s is saturation vapor pressure at 1.5- to 2.5-m height (kPa), e_d is mean actual vapor pressure at 1.5- to 2.5-m height (kPa), Δ is slope of the vapor pressure–temperature curve (kPa °C^{-1}), γ is psychrometric constant (kPa °C^{-1}), C_n (K mm s^3 Mg^{-1} day^{-1}) is a numerator constant that changes with crop height and calculation type step, and C_d (s m^{-1}) is a denominator that changes with crop height and calculation time step. Coefficients and variables for daily and hourly time steps are shown in Table 9.1.

FAO 1991 (Smith 1991) reference ET and ASCE standardized reference ET for tall and short crop was applied in south Florida for 2007. The FAO 1991 and ASCE short-crop estimates are not that different (Fig. 9.1): 1,322 and 1,314 mm, respectively for the year. The estimates from ASCE tall-crop estimates are very high,

Table 9.1 Standardized ASCE Penman–Monteith terms in the standardized reference equation

Coefficient/variable	Short crop	Tall crop
C_n	900	1,600
C_d	0.34	0.38
Reference vegetation height, h	0.12 m	0.50 m
Height of T and RH measurements, z_h	1.5–2.5 m	1.5–2.5 m
Height corresponding to wind speed, z	2.0 m	2.0 m
Zero plane displacement height, d	0.08	0.08
Latent heat of vaporization, λ	2.45 MJ kg^{-1}	2.45 MJ kg^{-1}
Surface resistance, r_s	70 s m^{-1}	45 s m^{-1}

Extracted from Allen et al. (2005)

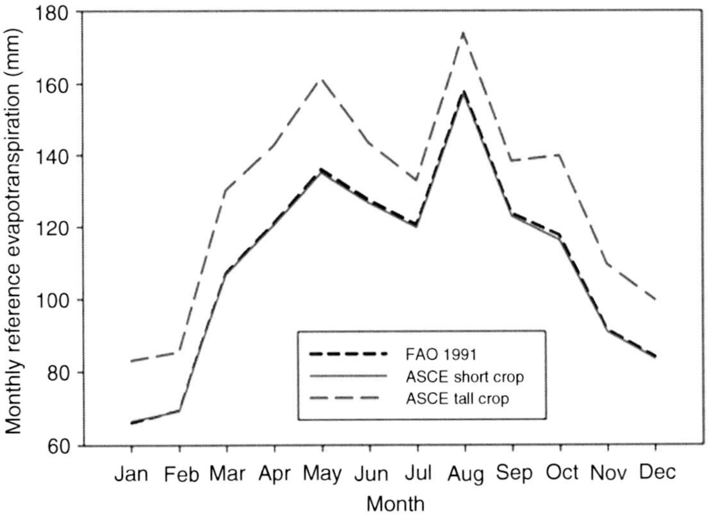

Fig. 9.1 Monthly reference evapotranspiration estimates by FAO 1991 and ASCE 2005 methods for a south Florida location in 2007

1,540 mm (Fig. 9.1). The reference ET estimates for short crop by the FAO 1991 and ASCE (2005) methods are the same as evaporation or potential evapotranspiration for the region at a weather station (26°38′N, 80°25′W) in south Florida (Abtew 2001; Abtew et al. 2003). The ASCE 2005 (Allen et al. 2005) estimates for tall crop are far higher than the potential evapotranspiration or evaporation in the area. From this analysis, it requires different sets of crop coefficient, K_c values for tall crops, to get reasonable ET estimates for south Florida and probably other locations too.

In a previous study, application of the FAO-24 or the FAO Blaney–Criddle method (Doorenbos and Pruitt 1977) reference evapotranspiration estimation gave an estimate of 1,384 mm year^{-1} in south Florida (Abtew and Sculley 1991).

9.4 Potential Evapotranspiration and Evaporation

Potential evapotranspiration is defined as the rate at which water, if available, is lost from wet surfaces and plant surfaces (Jensen et al. 1990). The distinction between potential evapotranspiration and reference crop evapotranspiration is stated that plant surfaces are not wet for reference evapotranspiration. Potential evaporation is defined as evaporation when water is not limiting and vapor pressure is saturated at the surface. A form of a Penman combination equation is given as Eq. 9.12 for estimating potential evaporation (E_{p}):

$$\lambda E_{\mathrm{p}} = \frac{\Delta}{\Delta + \gamma}(R_{\mathrm{n}} - G) + \frac{\rho c_{\mathrm{p}} k_{\mathrm{t}}}{\Delta + \gamma}(e_{\mathrm{a}} - e_{\mathrm{d}}) \tag{9.12}$$

where k_{t} is transfer coefficient and all other parameters are as defined earlier. Heat, mass, and momentum transfer coefficients and estimation equations are discussed in Chap. 4. Equation 9.12 gives the same estimates as the Penman combination Eq. 7.6 in Chap. 7 when the transfer coefficient (k_{t}) is 0.045 and other terms are the same.

9.5 Potential Evapotranspiration from Pan Evaporation

Pan evaporation (E_{pan}) has been used to estimate crop evapotranspiration by first estimating potential evapotranspiration ($\mathrm{ET_P}$) through a pan coefficient (K_{p}) and derive crop ET ($\mathrm{ET_c}$) through crop coefficient (K_{c}) as shown in Eqs. 9.13 and 9.14:

$$\mathrm{ET_p} = K_{\mathrm{p}} \times E_{\mathrm{pan}} \tag{9.13}$$

$$\mathrm{ET_c} = K_{\mathrm{c}} \times \mathrm{ET_p} \tag{9.14}$$

Pan coefficient (K_{p}) requires first estimating potential evapotranspiration by another method and using the ratio of the potential ET estimates and pan evaporation to develop seasonally varying, generally monthly, pan coefficients.

9.6 Crop Coefficients

Most studies on evapotranspiration deal with potential and reference evapotranspiration where water is not a limiting factor. Measurements of actual crop evapotranspiration are limited as ET varies by climate, soil type, moisture availability, and crop type. A generalized crop coefficients varying with crop growth stage were reported by Wright (1982). Also seasonally varying specific crop coefficients were developed for barley, peas, sugar beets, potatoes, corn, beans, winter wheat, and alfalfa for

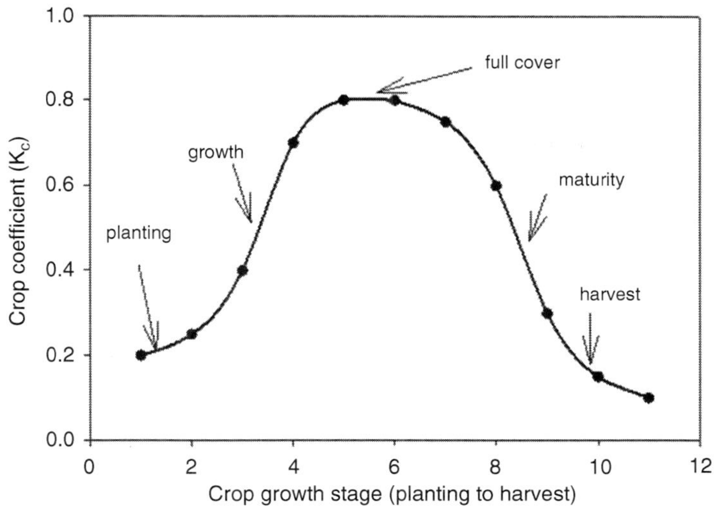

Fig. 9.2 General seasonal pattern for crop coefficient (K_c)

Kimberly, Idaho. The procedure used was to measure actual evapotranspiration with lysimeters and estimate reference evapotranspiration with a combination equation and meteorological data from the site. The ratio of actual ET to reference ET is the crop coefficient (Eq. 9.15):

$$K_c = \frac{ET_c}{ET_r} \tag{9.15}$$

From Eq. 9.15, it is apparent that, for a given crop ET_c, K_c values vary depending on the method used to derive reference evapotranspiration (ET_r). Crop coefficients have no transferable or applicable value unless the method of reference evapotranspiration is also stated. Crop coefficients of various crops, fruit trees, and other plants are given in FAO-24 along with reference ET estimation methods (Doorenbos and Pruitt 1977). ASCE Manual 70 (Jensen et al. 1990) provides K_c values for a number of crops for few regions. Figure 9.2 depicts a general pattern of K_c seasonal variation with crop growth for illustration purposes.

9.7 Summary

Estimating actual crop evapotranspiration will always be a challenge. Most studies on evapotranspiration are on potential and reference evapotranspiration where water is not a limiting factor. The standard for reference crops has been well-watered and clipped grass or alfalfa. Climate, soil type, location, and seasonal water limitation determine a specific crop's actual evapotranspiration. Efforts have been made in standardization of reference evapotranspiration estimation. There are various equations for estimating potential evapotranspiration. Crop evaporation studies are needed to develop crop coefficients. Otherwise, actual crop evapotranspiration will always be a gross estimate.

References

Abtew W (2001) Evaporation estimation for Lake Okeechobee in South Florida. J Irrig Drain 127(3):140–147

Abtew W, Sculley S (1991) Evapotranspiration estimation method for south Florida (The Everglades Agricultural Area). Paper presented at the 51st annual meeting of the Soil and Crop Science Society of Florida, Orlando, FL, 25–27 Sept 1991

Abtew W, Gregory JM, Borrelli J (1989) Wind profile: estimation of displacement height and aerodynamic roughness. Trans ASAE 32(2):521–527

Abtew W, Obeysekera J, Ortiz MI, Lyons D, Reardon A (2003) Evapotranspiration estimation for South Florida. In: Bizier P, DeBarry P (eds) Proceedings of the world water and environmental congress 2003. ASCE, Reston

Allen RG, Jensen ME (1989) Operational estimates of reference evapotranspiration. Agron J 81:650–662

Allen RG, Jensen ME, Wright JL, Burman RD (1989) Operational estimation of reference evapotranspiration. Agron J 81:650–662

Allen RG, Walter IA, Elliott R, Howell T, Itenfsu D, Jensen M (2005) The ASCE standardized reference evapotranspiration equation. ASCE, Reston

Doorenbos J, Pruitt WO (1977) Guidelines for predicting crop water requirements. FAO irrigation and drainage paper 24. FAO, Rome

Jensen ME, Burman RD, Allen RG (1990) Evapotranspiration and irrigation water requirements. ASCE manuals and reports on engineering practice No. 70. ASCE, Reston

Smith M (1991) Report on the expert consultation on procedures for revision of FAO guidelines for prediction of crop water requirements. FAO, Rome

Wright JL (1982) New evapotranspiration crop coefficients. J Irrig Drain Div ASCE 108(IR2): 57–74

Chapter 10
Spatially Distributed Surface Energy Flux Modeling

Abstract This chapter discusses the various models utilizing remotely sensed data for spatially distributed surface energy flux estimation. It outlines the basic principles and the corresponding approaches for estimating the different components of the energy budget (net radiation, soil heat flux, heat, sensible heat, and latent heat) and land surface parameters (albedo, emissivity, NDVI, surface temperature). The models discussed are capable of utilizing radiative and reflective data from satellite images from sensors of Landsat TM and ETM+, MODIS, ASTER, and others with thermal bands. The six models discussed include Surface Energy Balance Algorithm for Land (SEBAL), Two-Source Energy Balance (TSEB), Surface Energy Balance System (SEBS), Simplified Surface Energy Balance Index (S-SEBI), Mapping Evapotranspiration at High Resolution using Internalized Calibration (METRIC), and Simplified Surface Energy Balance (SSEB).

Keywords SEBAL • TSEB • S-SEBI • METRIC • SSEB • SEBS • Surface energy flux

10.1 Introduction

Several environmental disciplines such as hydrology, meteorology, climate science, and agronomy require knowledge of the land surface energy fluxes and budget. Reliable maps of surface energy fluxes are important for assessing surface–atmosphere interactions and exchange of water and energy between the earth's surface and the near ground level atmosphere. Surface energy balance models simulate microscale energy exchange processes between the ground surface and the atmospheric layer near the ground level. These processes include radiative, sensible heat, latent heat, and subsurface heat exchange processes. Spatially distributed energy budget computation will require spatial data from sources like satellite imagery and models that parameterize and utilize the different model parameters in the simulation. Results from these models provide climatic and land surface

W. Abtew and A. Melesse, *Evaporation and Evapotranspiration: Measurements and Estimations*, DOI 10.1007/978-94-007-4737-1_10,

information such as surface temperature, albedo, emissivity, radiation, and heat fluxes related to particular surfaces.

Remote sensing-based energy flux and surface parameters from different vegetated and nonvegetated surfaces are studied by various researchers. Energy flux from agricultural field (Kustas 1990; Bastiaanssen 2000; Kustas et al. 2004; Melesse and Nangia 2005), wetlands (Loiselle et al. 2001; Mohamed et al. 2004; Oberg and Melesse 2006; Melesse et al. 2007), rangelands and other vegetated surfaces (Kustas et al. 1994, 2003; Kustas and Norman 1999; French et al. 2000; Hemakumara et al. 2003; Melesse et al. 2008), lake evaporation (Melesse et al. 2009), and desert (Wang et al. 1998). These studies have shown the application of remote sensing in spatial mapping of flux and surface parameters to characterize the response of land surfaces to vegetation dynamics. Various flux surface energy balance models utilizing satellite imagery data are available. These models solve the very basic equation of the vertical energy budget where horizontal advections are assumed to be negligible and the net radiation is partitioned into latent heat, sensible heat, and soil heat flux. In this chapter, various spatially distributed energy budget models are presented and comparisons are made of model formulation and estimation of the different components of the energy budget.

The models discussed in this chapter include, Surface Energy Balance Algorithm for Land (SEBAL) (Bastiaanssen et al. 1998a, b), Two-Source Energy Balance (TSEB) (Norman et al. 1995a), Surface Energy Balance System (SEBS) (Jia et al. 2003), Mapping Evapotranspiration at High Resolution using Internalized Calibration (METRIC) (Allen et al. 2007), Simplified Surface Energy Balance (SSEB) (Senay et al. 2007), and Simplified Surface Energy Balance Index (S-SEBI) (Roerink et al. 2000).

Most of these models are similar except in some assumptions in estimating some of the model parameters in Eq. 10.1 below as well as sources of fluxes (single vs. two sources). The SSEB approach (Senay et al. 2007) is the most simplified of all which assumes the linear variation of the latent heat between the hot (minimum ET) and cold (maximum ET) pixels within the image assuming the temperature difference between soil surface and air is linearly related to soil moisture (Sadler et al. 2000).

Implementation of these models depends on the radiative, reflective, and thermal data from remote sensing mainly from sensors with thermal bands. Sensors commonly used for thermal mapping and surface energy flux estimation include the Landsat Thematic Mapper (TM), Landsat Thematic Mapper Plus (ETM+), ASTER (Advanced Spaceborne Thermal Emission and Reflection Radiometer), and MODIS (Moderate Resolution Imaging Spectroradiometer).

10.2 Remotely Sensed Data

10.2.1 Landsat

Since 1972, Landsat satellites have provided repetitive, synoptic, global coverage of high-resolution multispectral imagery. The Landsat TM instrument carried aboard

Table 10.1 Comparison of ASTER and Landsat ETM + sensors

	ASTER		ETM+	
Wavelength region	Band number (spatial resolution)	Spectral range (μm)	Band number (spatial resolution)	Spectral range (μm)
VNIR			1 (15 m)	0.45–0.52
	1 (15 m)	0.52–0.60	2 (30 m)	0.52–0.60
	2 (15 m)	0.63–0.69	3 (30 m)	0.63–0.69
	3 (15 m)	0.76–0.86	4 (30 m)	0.76–0.90
SWIR	4 (30 m)	1.60–1.70	5 (30 m)	1.55–1.75
	5 (30 m)	2.145–2.185	7 (30 m)	2.08–2.35
	6 (30 m)	2.185–2.225		
	7 (30 m)	2.235–2.285		
	8 (30 m)	2.295–2.365		
	9 (30 m)	2.360–2.430		
TIR	10 (90 m)	8.125–8.475	6 (60 m)	10.4–12.5
	11 (90 m)	8.475–8.825		
	12 (90 m)	8.925–9.275		
	13 (90 m)	10.25–10.95		
	14 (90 m)	10.95–11.65		

Landsat 4 and 5 (1982–present) is designed to achieve 30-m image resolution in seven spectral bands (Table 10.1). The Landsat ETM + instrument, carried aboard Landsat 7 (1999–present), includes new features that make it a more versatile and efficient instrument for global change studies, land-cover monitoring, and large area mapping than TM (Table 10.1). It has an enhanced sensor with a broad spectrum including a 15-m panchromatic and a 60-m by 60-m spatial resolution of the thermal band. Radiance data from sensors on satellites provide valuable information of watershed cover from which the thermal response of the surface, type, and extent of watershed cover can be easily determined. On May 31, 2003, the Scan Line Corrector (SLC) in the ETM + instrument failed. Without the effects of the SLC, the instrument images the Earth in a "zigzag" fashion, resulting in some areas that are imaged twice and others that are not imaged at all. The net effect is that approximately one-fourth of the data in a Landsat 7 scene is missing when acquired without a functional SLC. Landsat 7 continues to acquire data in this mode. Data products are available with the missing data optionally filled in using other Landsat 7 data selected by the user.

10.2.2 ASTER

ASTER (Advanced Spaceborne Thermal Emission and Reflection Radiometer) is one of the five state-of-the-art instrument sensor systems onboard Terra with a unique combination of wide spectral coverage and high spatial resolution in the visible near-infrared through shortwave infrared to the thermal infrared regions.

ASTER has 15-m resolution in three visible near-infrared (VNIR, 0.52–0.86 mm) bands and 90-m resolution in five thermal infrared (TIR, 8.1–11.6 mm) bands (Table 10.1).

10.2.3 MODIS

MODIS (Moderate Resolution Imaging Spectroradiometer) aboard Terra and Aqua satellites with its sweeping 2,330-km-wide viewing swath collects data from every point of the Earth's surface every 1–2 days in 36 discrete spectral bands. The spatial resolutions of MODIS bands are 250 m (bands 1–2), 500 m (bands 3–7), and 1,000 m (bands 8–36). They are designed to provide measurements in large-scale global dynamics including changes in Earth's cloud cover, radiation budget, and processes occurring in the oceans, on land, and in the lower atmosphere.

10.3 Surface Energy Budget and Models

In the absence of horizontally advective energy, the surface energy budget of land surface satisfying the law of conservation of energy can be expressed as

$$R_{\mathrm{n}} = \mathrm{LE} + H + G \tag{10.1}$$

where LE is latent heat or moisture flux (ET in energy units), R_{n} is net radiation at the surface, H is sensible heat flux to the air, and G is soil heat flux with a common unit for all parameters (W m^{-2}). Energy flux models solve Eq. 10.1 by estimating the different components separately.

10.3.1 Surface Energy Balance Algorithm for Land (SEBAL)

Surface Energy Balance Algorithm for Land (SEBAL) (Bastiaanssen et al. 1998a, b, 2005; Bastiaanssen 2000) is a single-source model that solves the surface energy balance equation to estimate LE as a residual. It uses thermal infrared data from satellite imagery including Landsat Thematic Mapper (TM) and Thematic Mapper Plus (ETM+), ASTER, and MODIS for estimating surface temperature and other model parameters. It also estimates net radiation, soil heat flux, albedo, emissivity using reflectance, and radiance values from the remotely sensed data, mainly from the above sensors.

The net radiation (R_{n}) absorbed by the surface is the sum of the net short (solar)- and longwave (thermal) radiations and is given by

$$R_{\text{n}} = (R_{\text{S}} \downarrow - R_{\text{S}} \uparrow) + (R_{\text{L}} \downarrow - R_{\text{L}} \uparrow) \tag{10.2}$$

where $R_{\text{S}\downarrow}$ and $R_{\text{S}\uparrow}$ are the incoming and outgoing, or reflected shortwave radiations, and $R_{\text{L}\downarrow}$ and $R_{\text{L}\uparrow}$ are incoming and outgoing longwave radiations, respectively.

Shortwave Radiation: The net shortwave radiation (R_{Sn}), in Eq. 10.3, is the balance between $R_{\text{S}}\downarrow$ and the $R_{\text{S}}\uparrow$:

$$R_{\text{Sn}} = R_{\text{S}} \downarrow - R_{\text{S}} \uparrow \tag{10.3}$$

SEBAL estimates $R_{\text{S}}\downarrow$ using

$$R_{\text{S}} \downarrow = \frac{G_{\text{sc}}\tau_{\text{sw}}}{\sin\theta {d_{e-s}}^2} \tag{10.4}$$

where G_{sc} is the solar constant expressed as 1,367 W m^{-2}, θ is solar inclination angle in radians, d_{e-s} is the relative distance between Earth and Sun in astronomical units, and τ_{sw} is one-way atmospheric transitivity, computed as a function of elevation (FAO-56) (Allen et al. 1998),

$$\tau_{\text{sw}} = 0.75 + 0.00002z \tag{10.5}$$

where z is the elevation above sea level (m) determined from the DEM (Digital Elevation Model). For a given albedo (α) or absorptivity (ζ_{short}), $R_{\text{S}\uparrow}$ in Eq. 10.3 can be derived based on $R_{\text{S}\downarrow}$:

$$R_{\text{S}} \uparrow = (1 - \zeta_{\text{short}}) R_{\text{S}} \downarrow = \alpha R_{\text{S}} \downarrow \tag{10.6}$$

In SEBAL, α of land-cover surfaces is computed using a relationship utilizing albedo of the top of the atmosphere (α_{toa}), estimated from the reflectance of remotely sensed data, the path radiance albedo ($\alpha_{\text{path radiance}}$), and τ_{sw} as

$$\alpha = \frac{\alpha_{\text{toa}} - \alpha_{\text{path - radiance}}}{\tau_{\text{sw}}^2} \tag{10.7}$$

The path radiance albedo ranges between 0.025 and 0.04, and Bastiaanssen (2000) recommends a value of 0.03.

Longwave Radiation: The net longwave radiation (R_{Ln}) is determined using $R_{\text{L}}\downarrow$ and $R_{\text{L}}\uparrow$:

$$R_{\text{Ln}} = R_{\text{L}} \downarrow - R_{\text{L}} \uparrow \tag{10.8}$$

The $R_{\text{L}\downarrow}$ is estimated using Eq. 10.9 as

$$R_{\text{L}} \downarrow = \sigma \varepsilon_{\text{a}} T_{\text{a}}^4 \tag{10.9}$$

where σ is the Stefan–Boltzmann constant (5.67×10^{-8} Wm^{-2} K^{-4}), ε_a is the atmospheric emissivity (dimensionless), and T_a is the near surface air temperature (K). The empirical equation for ε_a by Bastiaanssen et al. (1998a) is

$$\varepsilon_a = -0.85(\ln \tau_{sw})^{0.09} \quad (10.10)$$

$R_{L\uparrow}$ is the thermal radiation flux emitted from the Earth's surface to the atmosphere determined by the Stefan–Boltzmann law:

$$R_L \uparrow = \sigma \varepsilon_s T_{sur}^4 \quad (10.11)$$

where T_{sur} is the surface temperature (K) and ε_s is surface emissivity.

Thermal Infrared Surface Emissivity (ε_s)*:* Surface emissivity (ratio of the energy radiated by a surface to the energy radiated by a blackbody at the same temperature) is used to compute the surface temperature from thermal band of Landsat.

In SEBAL, ε_s is estimated using NDVI and an empirically derived method

$$\varepsilon_s = 1.009 + 0.047(\ln \text{NDVI}) \qquad (\text{NDVI} > 0) \quad (10.12)$$

Emissivity is assumed to be one for NDVI < 0.

Normalized Difference Vegetation Index: The NDVI (Rouse et al. 1974) is a measure of the degree of greenness in the vegetation cover of a land surface. It is the ratio of the difference to the sum of the reflectance values of NIR and red bands (Eq. 10.13):

$$\text{NDVI} = \frac{\text{NIR} - \text{RED}}{\text{NIR} + \text{RED}} \quad (10.13)$$

In highly vegetated areas, the NDVI typically ranges from 0.1 to 0.6, in proportion to the density and greenness of the plant vegetation. Clouds, water, and snow, which have larger visible reflectance than NIR reflectance, will yield negative NDVI values. Rock and bare soil areas have similar reflectance in the two bands and result in an NDVI near zero.

Surface Temperature: Surface temperature is an important parameter in understanding the exchange of energy between the Earth surface and the environment. Surface temperature is calculated from the thermal band radiance values of Landsat TM and ETM + sensors using the simplified Planck function (Eq. 10.14) (Markham and Barker 1986) and corrected using ε_s:

$$T_{sur} = \frac{K_2}{\ln\left(\frac{\varepsilon_s K_1}{R} + 1\right)} \quad (10.14)$$

where R is band 6 spectral radiance, ε_s is related to NDVI (Eq. 10.12), K_1 is calibration constant 1, and K_2 is calibration constant 2. For Landsat 5 TM, K_1 and

K_2 are 607.76 mWcm^{-2} sr^{-1} μm^{-1} and 1,260.56 K, respectively. For Landsat 7 ETM+, K_1 and K_2 are 666.09 Wm^{-2} sr^{-1} μm^{-1} and 1,282.71 K, respectively.

For Landsat 4/5 TM, R is a linear function of the digital number (DN):

$$R = m * \mathrm{DN} + d_{\mathrm{TM}} \tag{10.15}$$

where $m = 0.0056322$ mWcm^{-2} sr^{-1}μm^{-1} and $d_{\mathrm{TM}} = 0.1238$ mWcm^{-2} sr^{-1}μm^{-1}. R values for Landsat 7 ETM + were calculated as (NASA 2002)

$$R = \frac{(L_{\mathrm{max}} - L_{\mathrm{min}})}{254} * (\mathrm{DN} - 1) + L_{\mathrm{min}} \tag{10.16}$$

where L_{max} and L_{min} are maximum and minimum spectral radiance (Wm^{-2} sr^{-1} μm^{-1}). L_{max} and L_{min} are non-real-time postlaunch values, different for the low (6 L) and high (6 H) gain versions of the thermal band on ETM + .

Soil Heat Flux (G)*:* The soil heat flux (G) is the rate of heat storage to the ground from conduction. It is expressed in Campbell and Norman (1998) as

$$G = \lambda_s \frac{\mathrm{d}T}{\mathrm{d}z} \; \left(\mathrm{W\,m^{-2}}\right) \tag{10.17}$$

where λ_s is the thermal conductivity of the soil and dT is the difference in temperature along depth dz. Thermal conductivity (λ_s) of soil is a complex function of the conductivities and volume fractions of soil constituents, such as minerals, water, and air voids in the soil, as demonstrated in Campbell and Norman (1998). Field measurement of the above parameters is expensive and very difficult. Therefore, G is the most difficult one to measure, and empirical equations such as using the fraction of net radiation or relating it to vegetation cover are mostly used. SEBAL computes the ratio G/R_n using the following empirical equation representing values near midday (Bastiaanssen 2000):

$$\frac{G}{R_n} = 0.2(1 - 0.98\mathrm{NDVI}^4) \tag{10.18}$$

Sensible HeatFlux: Sensible heat flux (H) is estimated using the bulk aerodynamic resistance model and a procedure that assumes a linear relationship between the aerodynamic near-surface temperature–air temperature difference (dT) and T_s calculated from extreme pixels (cold and hot) to develop a linear relationship between dT and T_s. This approach avoids the need to know the air temperature as air temperature is assumed to be closer to the surface temperature of the colder pixel. This approach in SEBAL also assumes that at the colder and hotter pixel, H and LE are assumed to be zero, respectively, setting Eq. 10.1 at the colder pixel to be $R_n = G + \mathrm{LE}$ and, at the hotter pixel, $R_n = G + H$. LE is estimated as residual using Eq. 10.1. Figure 10.1 shows the daily spatial evapotranspiration map at the Glacial Ridge prairie restoration site in northwestern Minnesota estimated using SEBAL.

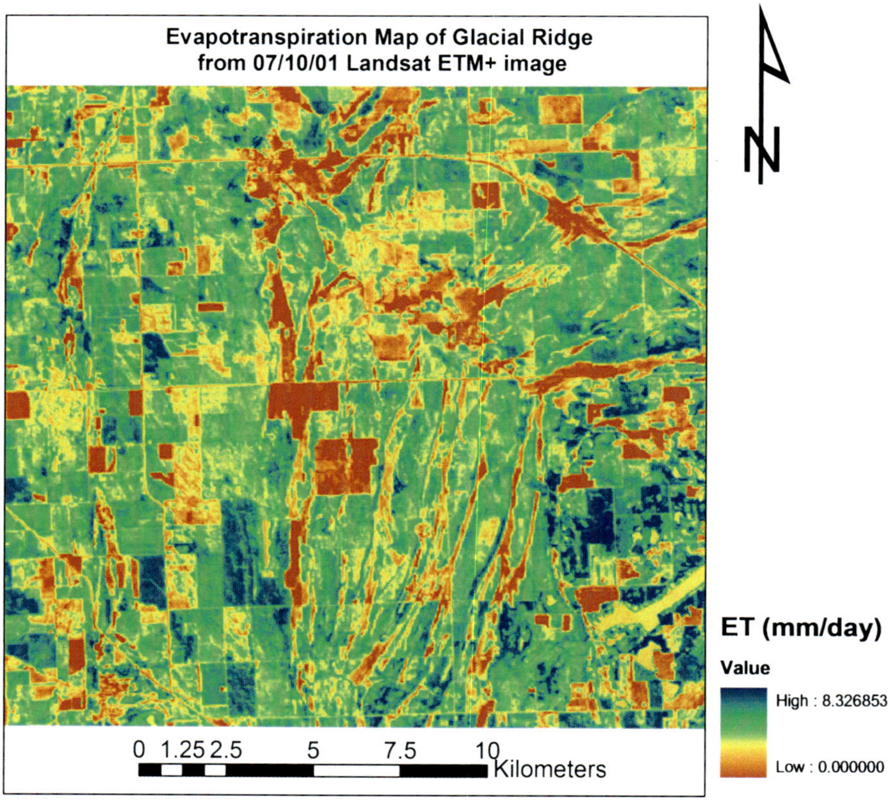

Fig. 10.1 Evapotranspiration map of the Glacial Ridge wetland restoration site (July 10, 2001), northwestern Minnesota from SEBAL

10.3.2 Two-Source Energy Balance (TSEB) Model

Two-source refers to the partition and treatment of an inhomogeneous land surface into two sources of heat and water vapor flux: the soil and the vegetation (Norman et al. 1995b). Vegetation and soil can have different energy and moisture exchanges with the overlying atmosphere, and fluxes from each source are assumed to be additive (Norman et al. 1995b).

The soil–vegetation system is approximated with a two-layer model, where the energy fluxes are partitioned between the soil and vegetation (Kustas 1990; Kustas and Norman 1999; French et al. 2000; Shuttleworth and Gurney 1990; Massman 1992; Norman et al. 1995b). Partitioning the components of the surface energy balance equation (Eq. 10.1) into the following vegetation and soil components:

$$R_n = R_{n,v} + R_{n,s} \quad \text{Net radiation} \tag{10.19}$$

$$R_{n,v} - H_v - \text{LE}_v = 0 \quad \text{Vegetation energy budget} \tag{10.20}$$

$$R_{n,s} - H_s - \text{LE}_s - G = 0 \quad \text{Soil energy budget} \tag{10.21}$$

where the subscripts v and s refer to the vegetation and soil components of the system, respectively.

The sensible and latent heat fluxes are partitioned into vegetation and soil components as

$$H = H_v + H_s = \rho C_p \left(\frac{T_v - T_a}{r_a}\right) + \rho C_p \left(\frac{T_s - T_a}{r_a + R_s}\right) \tag{10.22}$$

$$\text{LE} = \text{LE}_v + \text{LE}_s \quad \text{Latent heat} \tag{10.23}$$

where ρ is air density (kg m^{-3}), C_p is the heat capacity of air (1,004 J kg^{-1} K^{-1}), r_a is aerodynamic resistance, R_s is resistance to heat flow in the boundary layer immediately above the soil surface, T_v and T_s are vegetation and soil surface temperatures (K), respectively. r_a and R_s are estimated using the procedure shown in the TSEB model (Norman et al. 1995b).

To obtain a solution using Eq. 10.22, T_{sur} is related to T_v and T_s by Eq. 10.24 as (French et al. 2000)

$$T_{sur}^4 = f T_v^4 + (1 - f) T_s^4 \quad \text{Radiometric surface temperature} \tag{10.24}$$

where f is the fractional vegetation cover related to NDVI (Eqs. 10.25 and 10.26).

Fractional Vegetation Cover: To understand the change in the vegetation cover for images of different scenes and dates, the scaled NDVI (NDVI$_s$) has been used by many researchers (Price 1987; Che and Price 1992; Carlson and Ripley 1997; Carlson and Arthur 2000):

$$\text{NDVI}_s = \frac{\text{NDVI} - \text{NDVI}_{low}}{\text{NDVI}_{high} - \text{NDVI}_{low}} \tag{10.25}$$

where NDVI$_{low}$ and NDVI$_{high}$ are values for bare soil and dense vegetation, respectively.

Carlson and Ripley (1997) found the relationship between f and NDVI$_s$ to be

$$f \approx (\text{NDVI}_s)^2 \tag{10.26}$$

In the two-source model, as described by Norman et al. (1995b), Kustas and Norman (1999), and French et al. (2000), an initial estimation of LE$_v$ is obtained using the Priestley–Taylor approximation (Priestley and Taylor 1972) relating it to R_n within the green portion of the vegetation as

$$LE_v = 1.26 R_{n,v}\left[\frac{\Delta}{\Delta+\gamma}\right] \quad (10.27)$$

where Δ is the slope of the saturation vapor pressure versus temperature curve and γ is the psychometric constant.

$R_{n,s}$ is estimated from R_n and f (Norman et al. 1995b; French et al. 2000) using the following equation:

$$R_{n,s} = R_n e^{(0.9(\ln(1-f)))} \quad (10.28)$$

Hence, $R_{n,v}$ is given by

$$R_{n,v} = R_n - R_{n,s} = R_n\left(1 - e^{(0.9(\ln(1-f)))}\right) \quad (10.29)$$

Applying the Priestley–Taylor equation, H_v can be approximated as

$$H_v = R_{n,v}\left(1 - 1.26\frac{\Delta}{\Delta+\gamma}\right) \quad (10.30)$$

From Eqs. 10.22 and 10.30, T_v can be computed as

$$T_v = \left[\frac{R_a}{\rho C_p} R_{n,v}\left(1 - 1.26\frac{\Delta}{\Delta+\gamma}\right)\right] + T_a \quad (10.31)$$

This will estimate the initial value of T_v, and T_s will be estimated from Eq. 10.24 using T_{sur} from the thermal band of remotely sensed data. Using the value of T_s, the initial value of H_s is estimated using Eq. 10.22.

Rearranging Eq. 10.21 will yield LE_s as

$$LE_s = R_{n,s} - H_s - G \quad (10.32)$$

If LE_s is negative, LE_s is set to zero as this might indicate a dry soil, H_s is computed from the above equation, and T_s and T_v are computed again from H_s and T_{sur} formulations, respectively.

The new H_v is computed from

$$H_v = \rho C_p\left(\frac{T_v - T_a}{r_a}\right) \quad (10.33)$$

The new LE_v is computed from

$$LE_v = R_{n,v} - H_v \quad (10.34)$$

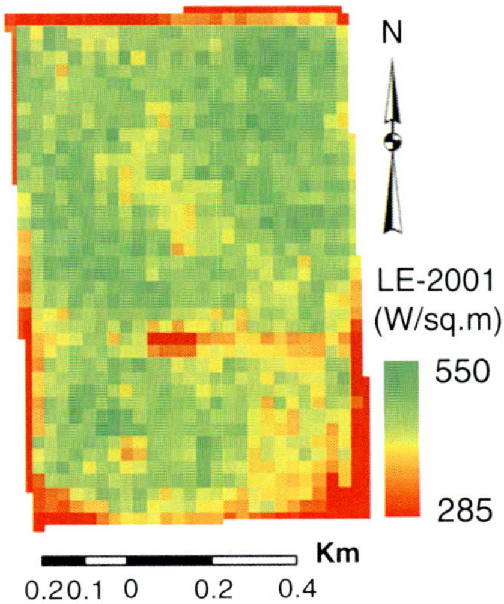

Fig. 10.2 Latent heat flux map of a soybean field, North Dakota from SEBAL–TSEB

Since Eq. 10.34 overrides the previous approximation of LE_v using Eq. 10.27, if LE_v is negative, LE_v is set to zero as this indicates a bare surface or stressed vegetation. New values of H_v, T_v, T_s, and H_s are computed from the respective equations. LE will be computed using Eq. 10.23. Figure 10.2 shows latent heat map of a soybean field in North Dakota mapped using the SEBAL–TSEB model.

The TSEB model has been applied to various agricultural fields including central and southern Arizona, the Sahel region of West Africa, Kansas, and central Oklahoma and has been found to be successful in modeling surface energy fluxes (Norman et al. 1995a; Kustas and Norman 1999; Schmugge et al. 1998, 2002; Anderson et al. 1997).

10.3.3 Surface Energy Balance System (SEBS)

The Surface Energy Balance System is a single-source surface energy balance model, that estimates atmospheric turbulent fluxes and surface evaporative fraction from remotely sensed data without partitioning the energy fluxes into the vegetation canopy and the soil surface (Friedl 2002). Unlike the two-source models, where canopy and soil flux partitioning is used, a single-source model uses only one resistance and assumes that all surfaces can be represented by one effective temperature and humidity value. SEBS was applied in different parts of the world (Su 2002; McCabe and Wood 2006; Jia et al. 2003).

SEBS applies both Bulk Atmospheric Similarity (BAS) and the Monin–Obukhov atmospheric surface layer (ASL) similarity in the model to determine turbulent fluxes at the regional and local scales, respectively (Su 2002). In SEBS, roughness height estimation takes into account surface heterogeneity.

SEBS requires three input data sets from different sources: (1) remotely sensed image-based inputs (albedo, emissivity, temperature, and the Normalized Difference Vegetation Index) to derive local surface roughness parameters; (2) meteorological parameters collected at a reference height (air pressure, temperature, relative humidity, wind speed); and (3) radiation data (downward solar radiation) and downward longwave radiation (Su 2002).

The SEBS model also consists of three modules: (1) energy balance estimation, (2) submodel to derive roughness length for heat transfer (Su et al. 2001), (3) submodel to derive stability parameters. Using these three modules, the energy balance for limiting boundary conditions (i.e., completely wet or dry pixels) can be resolved. This also allows the pixel-based derivation of the energy balance terms, relative evaporation, evaporative fraction, and evapotranspiration flux (Su 2002).

Net radiation is estimated based on the relationship shown in Eq. 10.2; the soil heat flux is estimated using Eq. 10.35:

$$G = R_n\left(\Gamma_c + (1 - f_c).(\Gamma_s - \Gamma_c)\right) \tag{10.35}$$

where Γ_c (G/R_n) for vegetation canopy and Γ_s (G/R_n) for soil are estimated to be 0.05 (Monteith 1973) and 0.315 (Kustas and Daughtry 1990), respectively. An interpolation is then performed between these limiting cases using the fractional canopy coverage, f_c, solved using Eq. 10.26.

In the atmospheric surface layer, the mean wind speed, u, and the mean temperature $\theta_o - \theta_a$ similarity can be written in integral form as

$$u = \frac{u_*}{k}\left[\ln\left(\frac{z-d}{z_{om}}\right) - \psi_m\left(\frac{z-d}{L}\right) + \psi_m\left(\frac{z_{om}}{L}\right)\right] \tag{10.36}$$

$$\theta_o - \theta_a = \frac{u_* H}{ku^* \rho C_p}\left[ln\left(\frac{z-d}{z_{oh}}\right) - \psi_m\left(\frac{z-d}{L}\right) + \psi_m\left(\frac{z_{oh}}{L}\right)\right] \tag{10.37}$$

where z is reference height, $u_* = (\tau_{o/\rho})^{1/2}$ is the friction velocity, τ_o is the surface shear stress, θ_o is the potential temperature at the surface, d is zero plane displacement, z_{om} is roughness height for momentum transfer, z_{oh} is roughness height for heat transfer, k is von Karman constant (0.4), ρ is density of air, c_p is specific capacity of air, θ_a is the potential air temperature at height z, ψ_m and ψ_h are the stability correction functions for momentum and sensible heat transfer, respectively, and L is the Obukhov length and is defined as

$$L = -\frac{\rho C_p u_*^3 \theta_v}{kgH} \tag{10.38}$$

where g is the acceleration due to gravity ($m^2\ s^{-1}$) and θ_v is the potential virtual temperature (K) near the surface.

Under the dry condition, the latent heat (or the evaporation) becomes zero due to the limitation of soil moisture, and the sensible heat flux is at its maximum value:

$$\mathrm{LE}_{\mathrm{dry}} = R_{\mathrm{n}} - H_{\mathrm{dry}} - G = 0 \tag{10.39}$$

$$H_{\mathrm{dry}} = R_{\mathrm{n}} - G \tag{10.40}$$

Under the wet condition, when the evaporation takes place at potential rate, λE_{wet}, (i.e., the evaporation is limited only by the energy available under the given surface and atmospheric conditions), the sensible heat flux takes its minimum value, H_{wet}, i.e.,

$$\lambda E_{\mathrm{wet}} = R_{\mathrm{n}} - H_{\mathrm{wet}} - G \tag{10.41}$$

$$H_{\mathrm{wet}} = R_{\mathrm{n}} - G - \lambda E_{\mathrm{wet}} \tag{10.42}$$

The relative evaporative fraction (Λ_{r}) is given by

$$\Lambda_{\mathrm{r}} = 1 - \frac{H - H_{\mathrm{wet}}}{H_{\mathrm{dry}} - H_{\mathrm{wet}}} \tag{10.43}$$

Evaporative fraction (Λ) is computed using

$$\Lambda = \frac{\lambda E}{R_{\mathrm{n}} - G} = \frac{\Lambda_{\mathrm{r}} * \lambda E_{\mathrm{wet}}}{R_{\mathrm{n}} - G} \tag{10.44}$$

$$H = (1 - \Lambda).(R_{\mathrm{n}} - G) \tag{10.45}$$

$$\lambda E = \Lambda(R_{\mathrm{n}} - G) \tag{10.46}$$

The derivation of the roughness length for heat transfer can be based on field estimates and literature values. In the absence of data, empirical relationships with NDVI for surface aerodynamic properties are employed. The empirical relation between the roughness length of momentum transfer, z_{om} (m), and NDVI used in this implementation of SEBS is given by (Su et al. 2001)

$$z_{\mathrm{om}} = 0.005 + 0.5\left(\frac{\mathrm{NDVI}}{\mathrm{NDVI}_{\mathrm{max}}}\right)^{2.5} \tag{10.47}$$

where $\mathrm{NDVI}_{\mathrm{max}}$ is the maximum NDVI within the image.

Canopy height, h_{c}, is derived using formulation from Brutsaert (1982) using the zero displacement (d_{o}):

$$h_c = \frac{3d_o}{2} \tag{10.48}$$

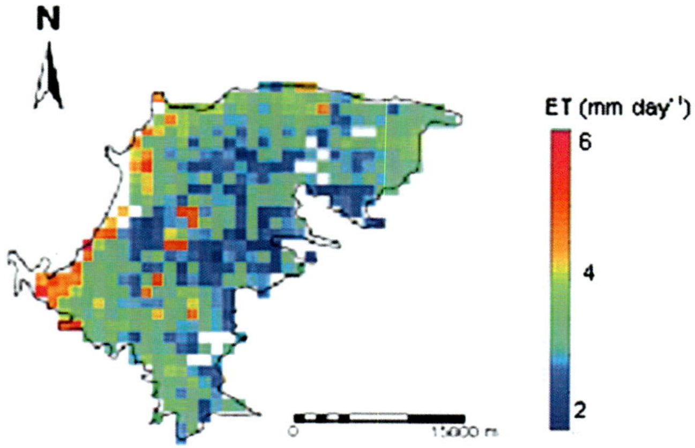

Fig. 10.3 Daily evapotranspiration of the Fogera floodplain, Ethiopia (April 26, 2008) using SEBS (Enku et al. 2011)

The details on estimating stability parameters in SEBS are given in Su et al. (2001). Figure 10.3 shows the daily evapotranspiration map of the Fogera floodplain using the SEBS model (Enku et al. 2011).

10.3.4 Mapping Evapotranspiration at High Resolution with Internalized Calibration (METRIC)

METRIC (Allen et al. 2007) has a similar approach to SEBAL for estimating the near-surface temperature gradient (dT), as an indexed function of radiometric surface temperature, which eliminates the need for accurate surface and air temperature measurements (Allen et al. 2007).

METRIC uses the concept of SEBAL in solving the surface energy balance equation. In METRIC, internal calibration of the surface energy balance is done at two extreme conditions (dry and wet) using locally available weather data. Both methods assume that the temperature difference between the land surface and the air (near-surface temperature difference) (dT) varies linearly with land surface temperature.

This relationship is derived based on two extreme pixels (*hot* and *cold*), representing dry and bare agricultural fields and wet and well-vegetated fields, respectively, in the images. Both SEBAL and METRIC methods use the linear relationship between the near-surface temperature difference and the land surface temperature (T_s) to estimate the sensible heat flux by assuming ET = 0 at the *hot* pixel, whereas at the *cold* pixel, maximum ET is assumed.

The first difference between SEBAL and METRIC is on the assumption of H and LE at the wet and dry pixels, respectively. Unlike SEBAL, METRIC does not assume $H = 0$ or $\mathrm{LE} = R_\mathrm{n} - G$ at the wet pixel. For the hot pixel, ET calculation is performed by solving the soil water budget, using meteorological data from a nearby weather station. METRIC also assumes for the wet pixel, $\mathrm{LE} = 1.05\ \mathrm{ET_r}\lambda_\mathrm{v}$, where $\mathrm{ET_r}$ is the tall-crop (e.g., alfalfa) reference ET calculated using the standardized ASCE Penman–Monteith equation applied using local meteorological observations. The second difference is that METRIC selects extreme pixels purely in an agricultural setting, where particularly the *cold* pixel needs to have biophysical characteristics (h_c, LAI) similar to the reference crop (alfalfa). The third difference is that METRIC uses the alfalfa reference evapotranspiration fraction ($\mathrm{ET_r}$ F) mechanism to extrapolate instantaneous LE flux to daily ET rates instead of using the Λ. The $\mathrm{ET_r}$ F is the ratio of $\mathrm{ET_i}$ (remotely sensed instantaneous ET) to the reference $\mathrm{ET_r}$ that is computed from weather station data at satellite overpass time. The benefits of using $\mathrm{ET_r}$ are the calibration around biases in R_n and G estimates at both ends of the temperature range:

$$\mathrm{ET_rF} = \frac{\mathrm{ET_i}}{\mathrm{ET_r}} \tag{10.49}$$

The fourth difference is that albedo in METRIC is estimated following Tasumi et al. (2008) to improve accuracy for a wide range of surface conditions. An additional benefit of using $\mathrm{ET_r}$ and $\mathrm{ET_r}$ F is the ability to account for general advection impacts on ET. Disadvantages are the requirement for relatively high-quality weather data on an hourly or shorter time step, and a shortcoming of METRIC is reliance on the accuracy of the $\mathrm{ET_r}$ estimate (Allen et al. 2007).

10.3.5 Simplified Surface Energy Balance (SSEB)

SSEB (Senay et al. 2007) assumes that the latent heat flux (actual evapotranspiration) varies linearly between the *hot* and *cold* pixels, similar to the principle of the SEBAL and METRIC in which sensible heat is assumed to vary linearly with the near-surface temperature difference. This assumption is based on the logic that the temperature difference between soil surface and air is linearly related to soil moisture (Sadler et al. 2000). On the other hand, crop soil water balance methods estimate actual ET using a linear reduction from the potential ET depending on soil moisture (Allen et al. 1998; Senay and Verdin 2003).

SSEB assumes that since *hot* pixels experience very little ET and *cold* pixels represent maximum ET throughout the study area, the average temperature of *hot* and *cold* pixels could be used to calculate proportional fractions of ET on a per pixel basis. The ET fraction ($\mathrm{ET_{frac}}$) is calculated for each pixel by applying the equation (Eq. 10.50)

$$\mathrm{ET_{frac}} = \frac{T_\mathrm{h} - T_\mathrm{x}}{T_\mathrm{h} - T_\mathrm{c}} \tag{10.50}$$

where T_h is the average of the three *hot* pixels selected for a given scene, T_c is the average of the three *cold* pixels selected for that scene, and T_x is the land surface temperature value for any pixel in the area.

The ET_{frac} is used in conjunction with a reference ET to calculate the per pixel actual ET values in a given image. The calculation of actual ET (AET) is achieved by Eq. 10.51:

$$AET = ET_{frac} * ET_{ref} \tag{10.51}$$

This simplified energy balance approach allowed the use of known reference ET at a coarse spatial resolution to derive spatially distributed ET measurements based on land surface temperature variability at 1-km resolution. Improvements in the spatial representation of ET distribution during the growing season can provide important insight into the extent of irrigated cropland, the quality of the growing season, and associated seasonal water use.

10.3.6 Simplified Surface Energy Balance Index (S-SEBI)

A Simplified Surface Energy Balance Index (S-SEBI) (Roerink et al. 2000) is a simplified method derived from SEBS to estimate surface fluxes from remote sensing data based on evaporative fraction and the contrast between the areas with extreme wet and dry temperature. The disadvantage of this method may be that it requires extreme T_s values (dry areas), which is not the case in every image. S-SEBI is a simpler method that does not need additional meteorological data and roughness length as in the case of SEBS.

10.4 Summary

For large area application, where point measurements are not practical, remote sensing-based surface energy flux modeling is very good for mapping spatially distributed water and energy fluxes. Various models are available and are tested in various regions of the world for different landscapes and ecosystems. These models have some limitations and are mostly applied to agricultural areas mainly for quantifying crop evapotranspiration for irrigation water management, crop water requirement estimation, crop yield estimation, and water budget estimation. Applications for different ecosystems such as grasslands, wetlands, forested areas, and savannas have been reported. The use of these models outside the ecosystem for which they were developed will necessitate the modification of the equations and validation of the results.

References

Allen RG, Pereira LS, Raes D, Smith M (1998) Crop evapotranspiration: guidelines for computing crop requirements. FAO irrigation and drainage paper No. 56. FAO, Rome

Allen RG, Tasumi M, Trezza R (2007) Satellite-based energy balance for mapping evapotranspiration with internalized calibration (METRIC) model. J Irrig Drain Eng 133(4):380–394

Anderson MC, Norman JM, Diak GR, Kustas WP, Mecikalski JR (1997) A two-source time integrated model for estimating surface fluxes using thermal infrared remote sensing. Remote Sens Environ 60:195–216

Bastiaanssen WGM (2000) SEBAL-based sensible and latent heat fluxes in the irrigated Gediz Basin, Turkey. J Hydrol 229:87–100

Bastiaanssen WGM, Menenti M, Feddes RA, Holtslag AAM (1998a) A remote sensing surface energy balance algorithm for land SEBAL. 1: formulation. J Hydrol 212–213:198–212

Bastiaanssen WGM, Pelgrum H, Wang J, Ma Y, Moreno J, Roerink GJ, van der Wal T (1998b) The surface energy balance algorithm for land SEBAL. 2: validation. J Hydrol 212–213:213–229

Bastiaanssen WGM, Noordman EJM, Pelgrum H, Davids G, Thoreson BP, Allen RG (2005) SEBAL model with remotely sensed data to improve water-resources management under actual field conditions. J Irrig Drain Eng 131(1):85–93

Brutsaert W (1982) Evaporation into the atmosphere. Reidel, Dordrecht

Campbell GS, Norman JM (1998) An introduction to environmental biophysics, 2nd edn. Springer, New York

Carlson TN, Arthur ST (2000) The impact of land use-land cover changes due to urbanization on surface microclimate and hydrology: a satellite perspective. Glob Planet Change 25:49–65

Carlson TN, Ripley AJ (1997) On the relationship between fractional vegetation cover, leaf area index and NDVI. Remote Sens Environ 62:24–252

Che N, Price JC (1992) Survey of radiometric calibration results and methods for visible and near-infrared channels of NOAA-7, -9 and -11 AVHRRs. Remote Sens Environ 41:19–27

Enku T, van der Tol C, Gieske ASM, Rientjes THM (2011) In: Melesse A (ed) Nile River Basin: hydrology, climate and water use. Springer, Dordrecht, p 163, 170

French AN, Schmugge TJ, Kustas WP (2000) Estimating surface fluxes over the SGP site with remotely sensed data. Phys Chem Earth 25(2):167–172

Friedl MA (2002) Forward and inverse modeling of land surface energy balance using surface temperature measurements. Remote Sens Environ 79:344–354

Hemakumara HM, Chandrapala L, Moene AF (2003) Evapotranspiration fluxes over mixed vegetation areas measured from large aperture scintillometer. Agric Water Manage 58(2): 109–122

Jia L, Zhongbo S, van den Hurk B, Menenti M, Arnold M, De Bruin Henk AR, Baselga YJJ, Manuel I, Antonio C (2003) Estimation of sensible heat flux using the Surface Energy Balance System (SEBS) and ATSR measurements. Phys Chem Earth A/B/C 28(1–3):75–88

Kustas WP (1990) Estimates of evapotranspiration with a one-and two-layer model of heat transfer over partial canopy cover. J Appl Meteorol 29:704–715

Kustas WP, Daughtry CST (1990) Estimation of the soil heat flux/net radiation ratio from spectral data. Agric For Meteorol 49(3):205–223

Kustas WP, Norman J (1999) Evaluation of soil and vegetation heat flux predictions using a simple two-source model with radiometric temperatures for partial canopy cover. Agric For Meteorol 94:13–29

Kustas WP, Perry EM, Doraiswamy PC, Moran MS (1994) Using satellite remote sensing to extrapolate evapotranspiration estimates in time and space over a semiarid rangeland basin. Remote Sens Environ 49(3):275–286

Kustas WP, Norman JM, Anderson MC, French AN (2003) Estimating sub-pixel surface temperatures and energy fluxes from the vegetation index–radiometric temperature relationship. Remote Sens Environ 85(4):429–440

Kustas WP, Li F, Jackson TJ, Prueger JH, MacPherson JI, Wolde M (2004) Effects of remote sensing pixel resolution on modeled energy flux variability of croplands in Iowa. Remote Sens Environ 92(4):535–547

Loiselle S, Bracchini L, Bonechi C, Rossi C (2001) Modeling energy fluxes in remote wetland ecosystems with the help of remote sensing. Ecol Model 45(2):243–261

Markham BL, Barker JL (1986) Landsat MSS and TM post calibration dynamic ranges, exoatmospheric reflectance, and at-satellite temperatures. EOSAT Landsat Tech Notes 1:3–8

Massman WJ (1992) A surface energy balance method for partitioning evapotranspiration data into plant and soil components for a surface partially canopy cover. Water Resour Res 28:1723–1732

McCabe MF, Wood EF (2006) Scale influences on the remote estimation of evapotranspiration, using multiple satellite sensors. Remote Sens Environ 105(4):271–285

Melesse A, Nangia V (2005) Spatially distributed surface energy flux estimation using remotely-sensed data from agricultural fields. Hydrol Process 19(14):2653–2670

Melesse A, Nangia V, Wang X, McClain M (2007) Wetland restoration response analysis using MODIS and groundwater data special issue: remote sensing of natural resources and the environment. Sensors 7:1916–1933

Melesse A, Frank A, Nangia V, Liebig M, Hanson J (2008) Analysis of energy fluxes and land surface parameters in grassland ecosystem: remote sensing perspective. Int J Remote Sens 29(11):3325–3341

Melesse A, Abtew W, Dessalegne T (2009) Evaporation estimation of Rift Valley Lakes in Ethiopia: comparison of models. Sensors 9(12):9603–9615. doi:10.3390/s91209603

Mohamed YA, Bastiaanssen WGM, Savenije HHG (2004) Spatial variability of evaporation and moisture storage in the swamps of the upper Nile studied by remote sensing techniques. J Hydrol 289:145–164

Monteith JL (1973) Principles of environmental physics. Edward Arnold, London

National Aeronautics and Space Administration (2002) Landsat 7 science data users handbook. Landsat Project Science Office, NASA Goddard Space Flight Center, Greenbelt, ver. November 8, 2002

Norman JM, Divakarla M, Goel NA (1995a) Algorithms for extracting information from remote thermal-IR observations of the earth's surface. Remote Sens Environ 51:157–168

Norman JM, Kustas WP, Humes KS (1995b) A two-source approach for estimating soil and vegetation energy fluxes from observations of directional radiometric surface temperature. Agric For Meteorol 77:263–293

Oberg J, Melesse AM (2006) Wetland evapotranspiration dynamics vs. ecohydrological restoration: an energy balance and remote sensing approach. J Am Water Resour Assoc 42(3):565–582

Price JC (1987) Calibration of satellite radiometers and the comparison of vegetation indices. Remote Sens Environ 21:15–27

Priestley CHB, Taylor RJ (1972) On the assessment of surface heat flux and evaporation using large-scale parameters. Mon Weather Rev 100:81–92

Roerink GJ, Su B, Menenti M (2000) S-SEBI: a simple remote sensing algorithm to estimate the surface energy balance. Phys Chem Earth B 25(2):147–157

Rouse JW, Haas RH, Schell JA, Deering DW (1974) Monitoring vegetation systems in the Great Plains with ERTS. In: Proceedings of the third earth resources technology satellite-1 symposium, NASA SP-351, Greenbelt, MD, pp 3010–3017

Sadler EJ, Bauer PJ, Busscher WJ, Millen JA (2000) Site-specific analysis of a droughted corn crop: water use and stress. Agron J 92:403–410

Schmugge TJ, Hook SJ, Coll C (1998) Recovering surface temperature and emissivity from thermal infrared multispectral data. Remote Sens Environ 65:121–131

Schmugge TJ, French AN, Kustas WP (2002) Evapotranspiration estimates using ASTER thermal infrared imagery. In: Owe M, D'Urso G (eds) Remote sensing for agriculture, ecosystems, and hydrology III, Proceedings of SPIE, vol 4542, Bellingham, WA, USA

Senay GB, Verdin J (2003) Characterization of yield reduction in Ethiopia using GIS-based crop water balance model. Can J Remote Sens 29(6):687–692

Senay GB, Budde M, Verdin JP, Melesse AM (2007) A coupled remote sensing and simplified energy balance approach to estimate actual evapotranspiration from irrigated fields special issue: remote sensing of natural resources and the environment. Sensor J 7:979–1000

Shuttleworth WJ, Gurney RJ (1990) The theoretical relationship between foliage temperature and canopy resistance in sparse crops. Q J R Meteorol Soc 116:497–519

Su Z (2002) The Surface Energy Balance System (SEBS) for estimation of turbulent heat fluxes. Hydrol Earth Syst Sci 6(1):85–99

Su Z, Schmugge T, Kustas W, Massman W (2001) An evaluation of two models for estimation of the roughness height for heat transfer between the land surface and the atmosphere. J Appl Meteorol 40:1933–1951

Tasumi M, Allen RG, Trezza R (2008) Estimation of at surface reflectance and albedo from satellite for routine, operational calculation of land surface energy balance. J Hydrol Eng 13(2):51

Wang J, Bastiaanssen WGM, Ma Y, Pelgrum H (1998) Aggregation of land surface parameters in the oasis-desert systems of Northwest China. Hydrol Process 12:2133–2147

Chapter 11
Crop Yield Estimation Using Remote Sensing and Surface Energy Flux Model

Abstract Spatial variability of energy fluxes calls for remote sensing-based approaches for mapping of fluxes, especially for larger areas. Cumulative consumptive use of water by crops can be related to crop yield with the help of remotely sensed data and surface energy balance models. This chapter discusses the use of surface energy balance models and Landsat images for correlating crop yield with latent heat flux. A case study for wheat and soybean fields is also presented. The modeling frame work and correlation of crop yield to spatially mapped latent heat, Bowen ratio, and wetness index is discussed. Net radiation was determined using the Surface Energy Balance Algorithm for Land (SEBAL) procedure. Applying the Two-Source Energy Balance (TSEB) model, surface temperature and latent and sensible heat fluxes were partitioned into vegetative and soil components and estimated at the pixel level. Results show that latent heat flux and Bowen ratio were correlated (positive and negative) to the yield data, respectively. The effect of microtopography on latent heat flux was shown using the wetness index and latent heat relationship. The flux estimation procedure from the SEBAL–TSEB model was useful and applicable to agricultural fields.

Keywords Remote sensing • Latent heat • SEBAL • Two-source model • Surface energy flux • Yield • Landsat • Bowen ratio • Wetness index • Microtopography

11.1 Introduction

Reliable maps of surface energy fluxes are important for assessing surface–atmosphere interactions. Surface energy balance models simulate microscale energy exchange processes between the ground surface and the near ground atmospheric layer level, and they are required by many environmental disciplines including

W. Abtew and A. Melesse, *Evaporation and Evapotranspiration: Measurements and Estimations*, DOI 10.1007/978-94-007-4737-1_11,

hydrology, agronomy, and meteorology. The energy exchange processes are highly spatiotemporal variable and include exchanges of radiative, sensible heat; latent heat; and subsurface heat. The high spatial variability of these fluxes and exchanges will limit point measurements for larger areas.

Methods for estimating energy fluxes from atmospheric measurements range from gross regional estimates to direct measurements of atmospheric gradients or fluxes. The former provides physical bounds over large areas, insensitive to local surfaces, and the latter measures at a single point (Tanner 1988). The estimation of surface energy fluxes using conventional ground-based procedures requires multiple measurements of variables controlling the process and is time, labor, and cost intensive. For large areas, remote sensing approaches are proven to be useful to estimate surface energy fluxes and parameters. Remote sensing provides data useful to estimate surface energy fluxes in the thermal infrared portion of the spectrum.

Remote sensing- and surface energy balance-based approaches are described in details in Chap. 10. One advantage of using remotely sensed data for hydrologic modeling and monitoring is its ability to generate information over large space and time scales, which is very useful for successful model analysis, prediction, and validation.

Surface energy fluxes are related to surface temperature, vegetative properties, and surface emissivities of land surface. For instance, low temperatures of land surface can be indicative of high moisture, irrigated field and/or vegetated cover, and hence latent heat dominance (Bowen ratio < 1). On the other hand, high temperatures can be an indication of the dominance of dry surface, low soil moisture or stressed vegetation, and hence higher sensible heat flux (Bowen ratio > 1). Vegetative properties can significantly affect the energy fluxes and exchange as they affect the surface air temperature gradient. This gradient determines fluxes of sensible heat and exchange of energy. Surfaces having different vegetation cover can have similar surface temperatures due to the difference in the aerodynamic properties of the surfaces (French et al. 2000).

Due to water vapor losses from agricultural fields, evapotranspiration may exhibit large spatial variability depending on the growth stage of crops and health of the vegetation; hence, studies focused at estimating spatial latent heat fluxes (evapotranspiration) of crop fields are very important. This knowledge can help in high-resolution irrigation water management practices.

The yield of many agricultural crops often can be predicted from the amount of water used by the crop in evapotranspiration (ET), which is the combined evaporation from the soil and transpiration by the crop (Hanks 1974). The relationship between yield and ET, called a crop water production function, has been widely and successfully used in various aspects of crop water management. The crop water production function is a mathematic model that reflects the rule of conversion between the crop yield and the water factor (ET). It is widely used in regional planning for improving irrigation efficiency and system evaluation. Based on the crop water production function, relationships between crop yield and consumptive water use have been developed. Such information can be used in crop water management to determine the amount of water that will result in the highest yield production per unit of water use (water use efficiency). The information from such analysis will be useful to farmers and irrigation managers, as well as to other researchers who investigate methods to increase the productivity and efficiency of crop water use.

11.2 Surface Energy Flux Budget

Surface energy flux estimation requires energy inputs, moisture conditions of soil and vegetation, and surface microclimate conditions (Norman et al. 1995a; French et al. 2000). Remote sensing has proven to provide the energy inputs (short- and longwave radiations) and surface moisture conditions of soil and vegetation (surface temperature and vegetation indices) at a reasonable spatial and temporal scale. Surface microclimate can be collected from networks of meteorological stations.

As shown in Chap. 10, in the absence of horizontally advective energy, the surface energy budget of land surface satisfying the law of conservation of energy can be expressed as

$$R_{\mathrm{n}} - \mathrm{LE} - H - G = 0 \tag{11.1}$$

Energy flux models solve Eq. 11.1 by estimating the different components separately. Remote sensing-based models have proven the ability to address the spatial variability of these fluxes by computing the value of energy budget components in the equation above at pixel level.

As discussed in Chap. 10, the Surface Energy Balance Algorithms for Land (SEBAL) (Bastiaanssen et al. 1998a, b) and the Two-Source Energy Balance (TSEB) (Norman et al. 1995b; Kustas and Norman 1999) models utilize remotely sensed data such as Landsat, ASTER, and MODIS to solve Eq. 11.1 by computing surface energy fluxes from satellite images and meteorological data.

The SEBAL model has been used in various studies to assess evapotranspiration rates in the USA, Spain, Italy, Turkey, Pakistan, India, Sri Lanka, Egypt, Niger, and China (Bastiaanssen et al. 1998a, b; Wang et al. 1998; Bastiaanssen 2000; Morse et al. 2000; Melesse and Nangia 2005; Melesse et al. 2007).

This chapter discusses the use of the coupled SEBAL–TSEB model and Landsat imagery in estimating latent heat fluxes from wheat and soybean agricultural fields and makes comparisons to actual yield of the crop fields. The two models are discussed in detail in Chap. 10.

In order to estimate daily evapotranspiration, instantaneous ET was converted to daily ET using the following equations:

$$\mathrm{ET}_{\mathrm{inst}} = 3600 \frac{\lambda \mathrm{ET}}{\lambda} \tag{11.2}$$

Table 11.1 Coefficients for ASCE-PM Ref-ET equations (Allen et al. 2005)

	Short reference (ET_o)		Short reference (ET_r)			
Computation time step	C_n	C_d	C_n	C_d	Units	Units for R_n, G
Daily or monthly	900	0.34	1,600	0.38	mm day^{-1}	MJ m^{-2} day^{-1}
Hourly – daytime	37	0.24	66	0.25	mm h^{-1}	MJ m^{-2} h^{-1}
Hourly – nighttime	37	0.96	66	1.7	mm h^{-1}	MJ m^{-2} h^{-1}

$$\mathrm{ET_r\,F} = \frac{\mathrm{ET_{inst}}}{\mathrm{ET_r}} \tag{11.3}$$

$$\mathrm{ET_{24}} = \mathrm{ET_r\,F} \times \mathrm{ET_{r24}} \tag{11.4}$$

ASCE-PM Standardized Reference ET equation (Jensen et al. 1990) is given by

$$\mathrm{ET_{ref}} = 0.408 \frac{\Delta(R_n - G) + \gamma \frac{C_n}{T+273} u_2 (e_s - e_d)}{\Delta + \gamma(1 + C_d u_2)} \tag{11.5}$$

where $\mathrm{ET_{ref}}$ is either the short ($\mathrm{ET_o}$) or tall ($\mathrm{ET_r}$) reference ET (mm day^{-1}, or mm h^{-1}), R_n is the net radiation at the crop surface (MJ m^{-2} day^{-1} or MJ m^{-2} h^{-1}), T is the mean daily or hourly temperature at a 1.5–2.5-m height (°C), G is the soil heat flux density at the soil surface (MJ m^{-2} day^{-1} or MJ m^{-2} h^{-1}), u_2 is the mean daily or hourly wind speed at a 2-m height (m s^{-1}), e_s is the mean actual saturation vapor pressure at 1.5–2.5-m height (kPa), e_d is the mean actual vapor pressure at 1.5–2.5-m height (kPa), Δ is the slope of the vapor pressure–temperature curve (kPa °C^{-1}), γ is the psychrometric constant (kPa °C^{-1}), and C_n is a function of the computation time step (hourly or daily) and of the aerodynamic resistance, which is a function of the reference type: grass or alfalfa. The term C_d is a constant that is a function of the surface resistance values, which are also functions of the reference type: grass or alfalfa. Jensen et al. (2000) gave the values of C_d and C_n as shown in Table 11.1.

11.3 Case Study

The surface energy flux versus crop yield study was conducted for six growing seasons from 1997 to 2002 on four contiguous fields located in Polk County, Northwestern Minnesota (Fig. 11.1). The study fields covering an area of 250 ha (2.5 km^2) are located in the Red River Valley, which is one of the nation's most fertile agricultural areas. Wheat and sugar beets are the most important crops in Polk County. Barley has the second most acreage but stands as the third most economically important crop. The study uses yield and ET data from wheat and soybean fields.

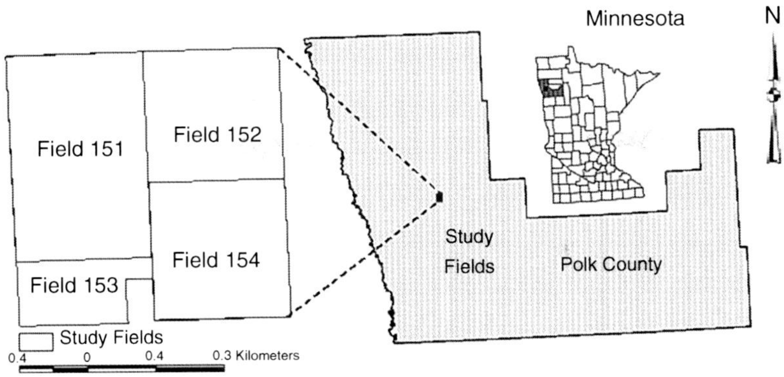

Fig. 11.1 Location of the study fields on the map of Minnesota

Table 11.2 Landsat images used in the study

Date	Sensor	No. of bands	Spatial resolution
June 21, 1997	TM	7	30-m (visible, NIR and MIR), 120-m (TIR)
July 10, 1998	TM	7	30-m (visible, NIR and MIR), 120-m (TIR)
July 21, 1999	TM	7	30-m (visible, NIR and MIR), 120-m (TIR)
July 23, 2000	ETM+	8	30-m (visible, NIR and MIR), 60-m (TIR) panchromatic (15-m)
July 10, 2001	ETM+	8	30-m (visible, NIR and MIR), 60-m (TIR) panchromatic (15-m)
July 13, 2002	ETM+	8	30-m (visible, NIR and MIR), 60-m (TIR) panchromatic (15-m)

The average winter temperature of the area is −13°C with the average daily minimum temperature of −18°C. In summer, the average daily temperature is 20°C with average maximum of 30°C. The total annual precipitation of the area is 505 mm with 70% of the precipitation occurring in the months of April through September. The growing season for most crops falls within this period. The soils in Polk County generally are dark and range in texture from clayey to sandy. Soils in the western half of the county were formed in silty and clayey lacustrine sediments.

11.3.1 Data

The study used remotely sensed data (Landsat Thematic Mapper, TM, and Enhanced Thematic Mapper Plus, ETM+), Digital Elevation Model (DEM), and crop data (yield) and weather data (solar radiation, wind speed, and air temperature).

Seven Landsat TM and ETM + images (Table 11.2) were used to process the intermediate parameters (Normalized Difference Vegetation Index, NDVI), fractional vegetation cover, radiometric surface temperature corrected using surface

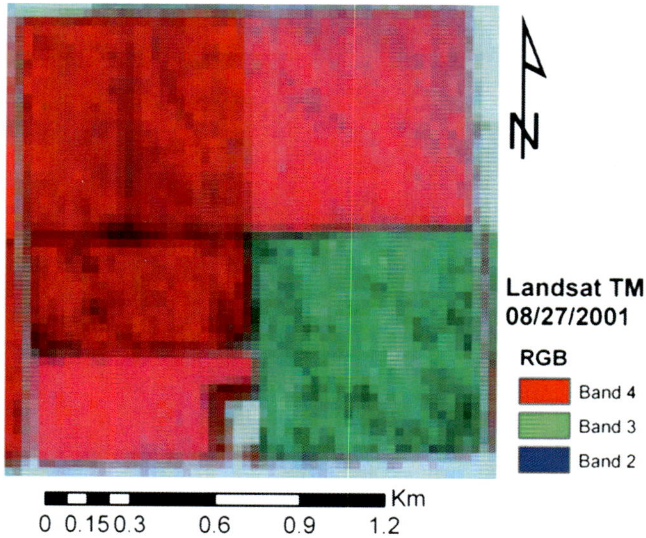

Fig. 11.2 Landsat images of the study area

Fig. 11.3 5-m Digital Elevation Model (DEM) of the field

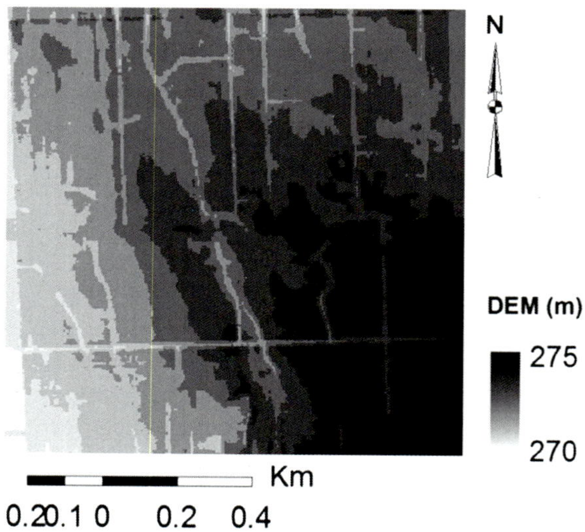

emissivity, albedo, and surface and atmospheric emissivities, from which surface energy flux components were estimated. The Landsat images were selected to represent the growing stage of the crops at full canopy. Figure 11.2 shows one of the Landsat images of the fields used in this analysis.

The 5-m DEM of the field, used in the energy flux computation, was mapped during land preparation (Fig. 11.3). The 5-m yield grids (bushels/acre) for each

Table 11.3 Yield data by season and field

Year	Field 151	Field 152	Field 153	Field 154
1997	Wheat			
1998		Wheat	Soybean	
1999	Wheat		Soybean	
2000		Wheat		Soybean
2001	Soybean			Wheat
2002	Wheat			

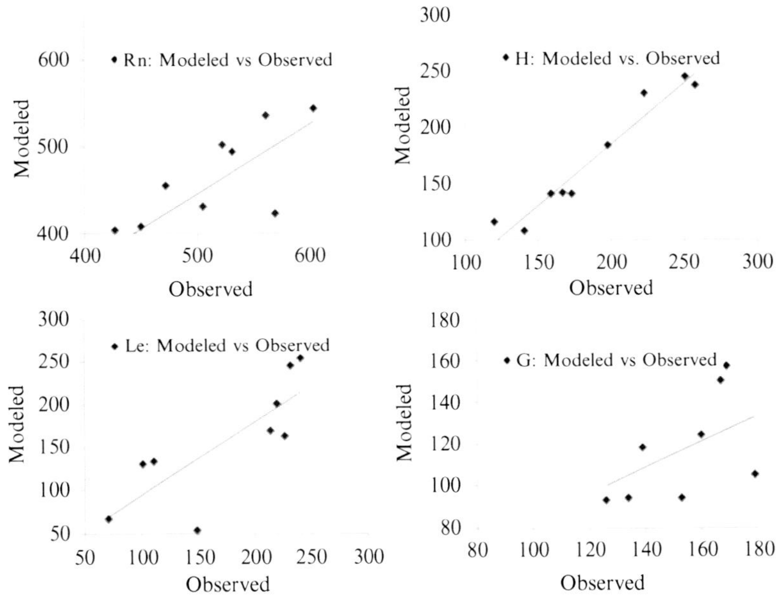

Fig. 11.4 Modeled vs. observed energy fluxes at the Fort Peck, Montana, site

season and field were developed from yield point data collected from the combine harvester's yield monitor. Table 11.3 shows the year and planted crop type by field. Yield data were correlated with the surface energy fluxes determined from the Landsat TM and ETM + sensors.

The weather data, which were used as input to the surface energy flux model, include wind speed and air temperature. Hourly values were collected from the weather station at the study field. Only values at the time of the Landsat overpass of the area were used.

Since the study fields did not have on-site flux measurements, calibration and validation of the modeled fluxes were done using data from the micrometeorology flux tower located at Fort Peck, Montana. Validation results and comparison of the modeled and observed fluxes are shown in Melesse and Nangia (2005). Figure 11.4 shows the comparison of the modeled and predicted fluxes using the SEBAL model at the Fort Peck, Montana, USA, experimental site.

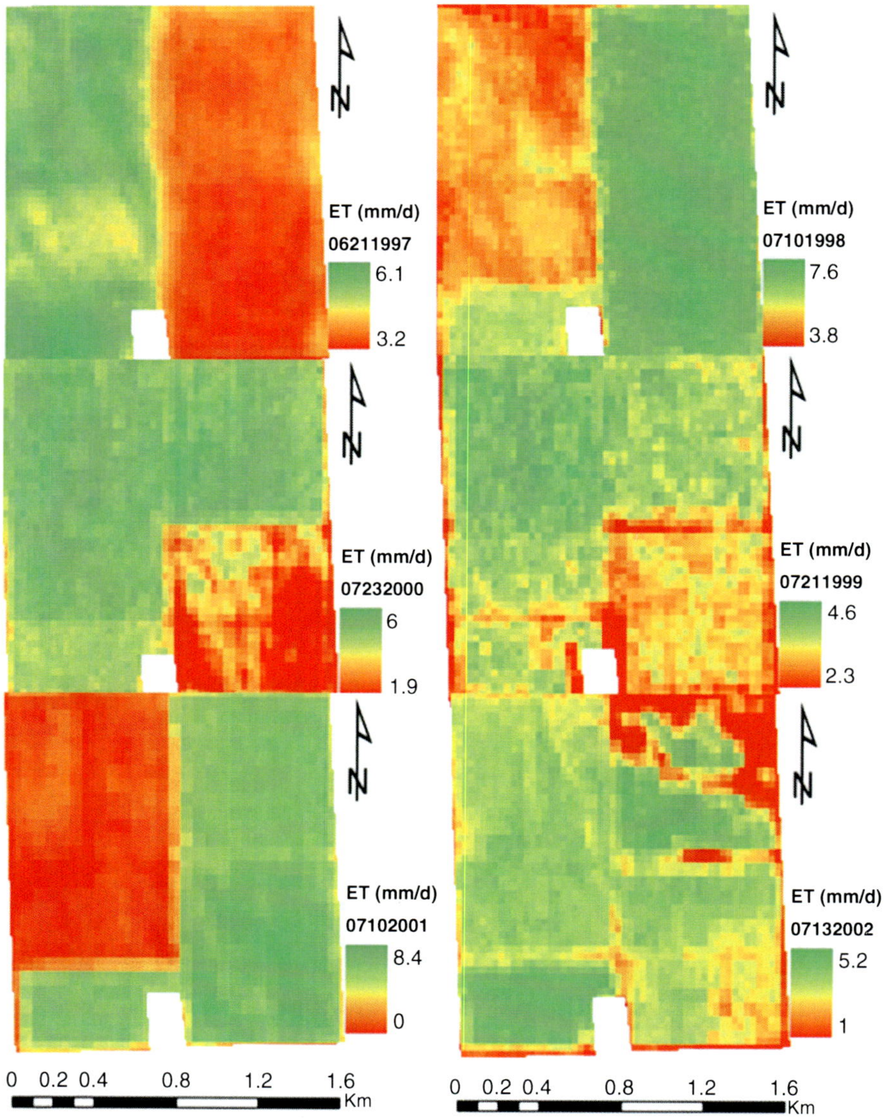

Fig. 11.5 Landsat-based spatial daily evapotranspiration of the study area (1997–2002)

11.3.2 Results and Discussion

Evapotranspiration: Spatial evapotranspiration maps from the six images (1997–2002) are shown in Fig. 11.5. These are daily ET values on the respective image dates. Since the different fields are planted different crops at the different periods, the variation in the spatial and temporal ET emanates from this variation.

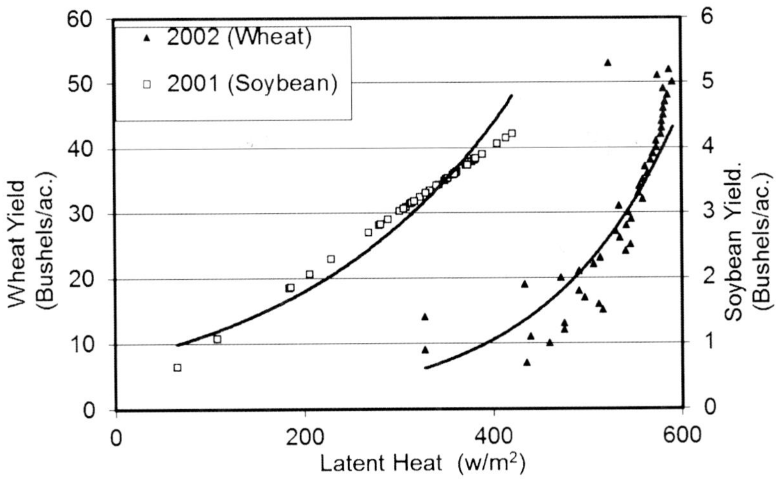

Fig. 11.6 Yield vs. total latent heat (wheat and soybean) from Field 151

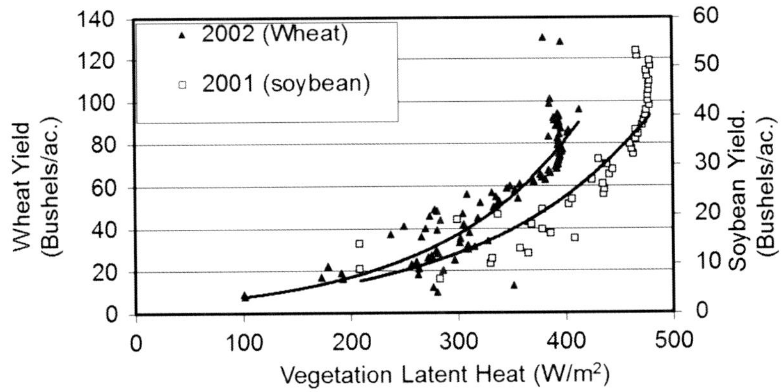

Fig. 11.7 Yield vs. vegetation latent heat (wheat and soybean) from Field 151

Evapotranspiration values in energy units (latent heat flux) were converted to grids and georeferenced to the respective yield grids. Latent heat grids were summarized using integer values of yield grids, and scattergrams were drawn using the mean values of LE for each value of yield. For instance, those latent heat pixels having a yield of 20 bushels/acre were identified, and their mean latent heat value was computed. Scattergrams were drawn using the data categorized by crop (wheat and soybean) and season (2001–2002) for Field 151.

The scattergrams (Figs. 11.6 and 11.7) show that crop yield increases exponentially with the increase of latent heat (total) and vegetative latent heat, with an average R^2 of 0.67 (wheat) and 0.70 (soybean).

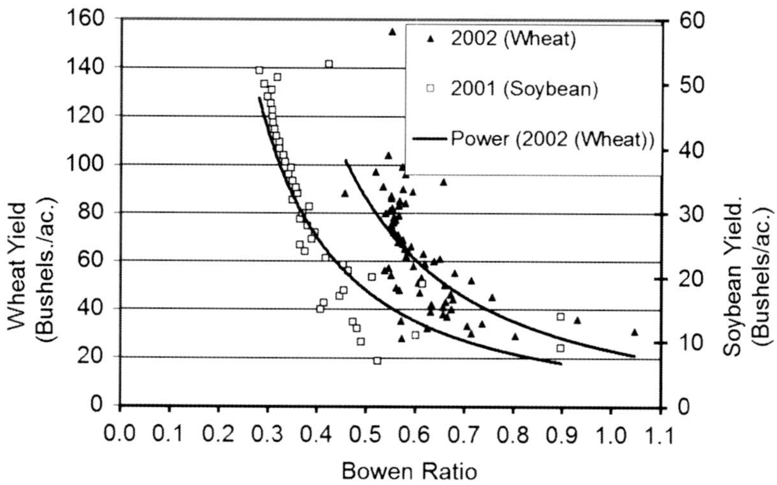

Fig. 11.8 Scattergram showing yield vs. Bowen ratio from Field 151

Yield Versus *Bowen Ratio(Field 151):* The Bowen ratio (B) (Bowen 1926) is computed as the ratio of H to LE. Bowen ratio shows the relative proportions of sensible and latent heat. Higher values of B ($B > 1$) indicate dominance of sensible heat, which is the case for dry soil or stressed vegetation with little evaporation from the soil and reduced respiration from the crop. On the other hand, lower values of B ($B < 1$) are indications of dominance of respiration and evaporation process over sensible heat loss from the soil and canopy to the air. This is typical of a wet soil and vegetated surface. Scattergrams of yield (Fig. 11.8) versus B for Field 151 in the 2001 and 2002 seasons show a negative correlation for both growing seasons, indicating dominance of latent heat flux over the sensible heat from the vegetative surfaces (Fig. 11.8).

Yield Versus *Wetness Index(WI) (Field 151):* Topography is a determinant for magnitudes and spatial distributions of water and energy fluxes over natural landscapes. The topographic configuration of a landscape is a control boundary condition for the hydrologic processes of surface runoff, evaporation, and infiltration, which take place at the ground–atmosphere interface. For example, wetness index (WI) provides a description of the spatial distribution of soil moisture in terms of topographic information. WI is computed as

$$\mathrm{WI} = \mathrm{Ln}\left(\frac{A}{S}\right) \tag{11.6}$$

where A and S are the specific drainage (i.e., flow accumulation) area and slope, respectively.

As A increases and/or S decreases, WI becomes larger, indicating that soil moisture content will increase. Because WI takes into account local slope variations,

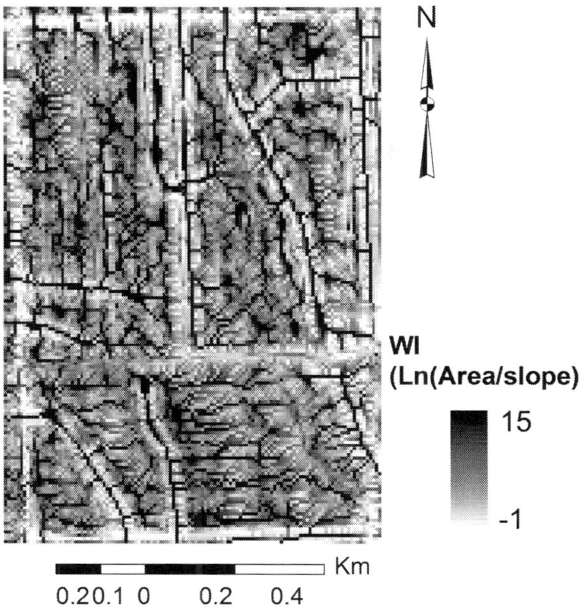

Fig. 11.9 Wetness index (*WI*) of the study area

it has proven to be a reasonable indicator for soil wetness, flow accumulation, saturation dynamic, water table fluctuation, evapotranspiration, soil horizon thickness, organic matter content, pH, silt and sand content, and plant cover density (Kulagina et al. 1995; Florinsky 2000). Wetness index (WI) for the study area is depicted in Fig. 11.9.

The microtopography expressed in the form of the wetness or topographic index and yield from wheat are found to be positively correlated in areas where WI values were low up to a certain extent (Fig. 11.10). This may be because the microtopography controls soil moisture content as well as its spatial distribution. Grids with higher WI values are identified as the areas receiving more overland flows (i.e., with greater flow accumulations) and having a smaller gradient. These areas have higher soil moisture but a higher evaporation rate than the areas with lower WI values. The correlation between WI and soil moisture is further verified by the observation that when water is a limiting factor of an agricultural field, the crop in the areas with higher WI values tends to grow better than the crop in the areas with lower WI values. This can be attributed to more water availability for transpiration (i.e., latent heat demand) in areas with higher WI values. Figure 11.10 shows WI versus yield for wheat and soybean for 2002 and 2001 seasons, respectively. The relation between yield and WI for soybean was not significant.

Yield Prediction ErrorAnalysis: Once the correlation between yield and LE was estimated, a predicted spatial map of yield was generated. In order to show the accuracy of the prediction, the residual mean and standard deviation were calculated (Tables 11.4 and 11.5). From the prediction error analysis, it is shown

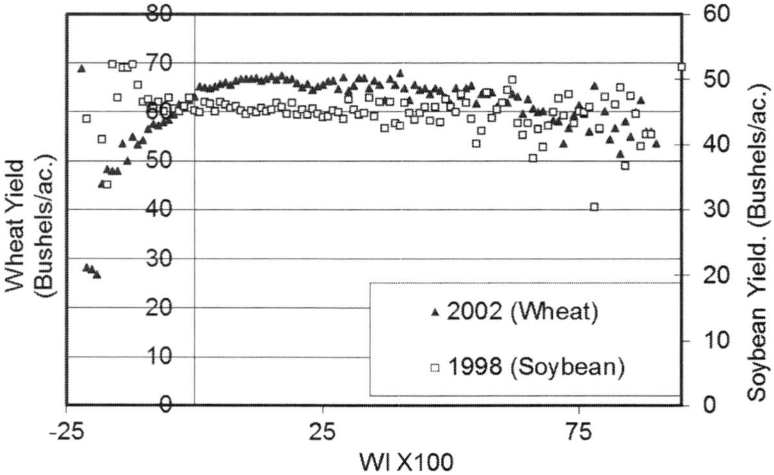

Fig. 11.10 Scattergram showing yield vs. wetness index (*WI*) from Field 151

that the average residual means for wheat and soybean fields were −4.2 and 0.11 bushels/acre, respectively. Similarly, the average standard deviations of the residuals for the wheat and soybean fields were 16.2 and 16.6 bushels/acre, respectively. The error analysis results show that the average error of prediction for soybean fields was smaller than that of the wheat fields. It is also shown that the average percentage of under- or overprediction (ratio of mean residual to observed mean) was also smaller for soybean (1.4%) when compared to wheat (−9.6%).

11.4 Summary

This chapter discussed the role of remotely sensed data and spatially distributed energy flux modeling in estimating evapotranspiration and hence correlating the same to crop yield. The demonstrated case study estimated spatial surface energy fluxes determined from remotely sensed data for seven growing seasons at four crop fields. Energy fluxes were calibrated and verified using flux tower data. Relationships between crop yield to LE, wetness index, and Bowen ratio were established. The SEBAL–TSEB model predicted components of the surface energy budget with reasonable accuracy.

The LE versus yield relationship was good, with an average R^2 of 0.67 for wheat and 0.70 for soybean, respectively. Similarly, the average mean errors of the predicted yield were −0.42 and 1.1 bushels/acre for wheat and soybean, respectively. For developing a better understanding of the close relationships between the LE and crop yield, more data from different fields, crops, and growing seasons will be helpful.

Table 11.4 Mean and standard deviation (SD) of residuals and percent under- or overprediction (% Pred.) of yield by field

	Field 151			Field 152			Field 153			Field 154		
	Residual			Residual			Residual			Residual		
Year	Mean (bu ac^{-1})[a]	SD (bu ac^{-1})	Pred. (%)	Mean (bu ac^{-1})	SD (bu ac^{-1})	Pred. (%)	Mean (bu ac^{-1})	SD (bu ac^{-1})	Pred. (%)	Mean (bu ac^{-1})	SD (bu ac^{-1}	Pred. (%)
1997	−1.99	9.8	−4.7	–	–	–	–	–	–	–	–	–
1998	–	–	–	−4.29	17.51	−8.9	12.68	11.2	37.1	–	–	–
1999	−1.4	8.6	−2.07	–	–	–	−6.73	16.9	−16.5	–	–	–
2000	–	–	–	−5.15	11.62	−23.7	–	–	–	−7.4	31.22	19.6
2001	1.87	7.03	4.77	–	–	–	–	–	–	4.77	28.7	−7
2002	−7.41	20.86	11.27	–	–	–	–	–	–	–	–	–
Average	−2.23	11.57	−3.37	−4.72	14.55	−16.3	2.98	14.1	10.3	6.09	29.96	13.3

[a] 1 bushel wheat/soybeans = 27.22 (27) kg and 1 bu/ac wheat/soybeans = 67.25 kg ha^{-1}

Table 11.5 Mean and standard deviation (SD) of residuals and percent under- or overprediction (% Pred.) of yield by crop

	Residuals (all fields)		
Crop	Mean (bu ac^{-1})	SD (bu ac^{-1})	Pred. (%)
Wheat	−4.17	16.19	9.60
Soybean	0.11	16.60	1.42

Acknowledgments The authors acknowledge George Seielstad, Gary Johnson, Ofer Beeri, Grant Casady, David Baumgartner, Santhosh Seelan, Jason Oberg, Chris Carlson, Ganesh Pulicherla, and other members of the Upper Midwest Aerospace Consortium for their help. The authors extend their appreciation to Gary Wagner for providing yield and other field data and Tilden Meyers of NOAA for providing flux tower data for the model validation.

References

Allen RG, Walter IA, Elliott R, Howell T, Itenfsu D, Jensen M (2005). The ASCE Standardized Reference Evapotranspiration Equation. ASCE, Reston, VA

Bastiaanssen WGM (2000) SEBAL-based sensible and latent heat fluxes in the irrigated Gediz Basin, Turkey. J Hydrol 229:87–100

Bastiaanssen WGM, Menenti M, Feddes RA, Holtslag AAM (1998a) A remote sensing surface energy balance algorithm for land SEBAL. 1: formulation. J Hydrol 212–213:198–212

Bastiaanssen WGM, Pelgrum H, Wang J, Ma Y, Moreno J, Roerink GJ, van der Wal T (1998b) The surface energy balance algorithm for land SEBAL. 2: validation. J Hydrol 212–213:213–229

Bowen IS (1926) The ratio of heat losses by conduction and evaporation from any water surface. Phys Rev 27:779–787

Florinsky IV (2000) Relationships between topographically expressed zones of flow accumulation and sites of fault intersection: analysis by means of digital terrain modeling. Environ Model Softw 15:87–100

French AN, Schmugge TJ, Kustas WP (2000) Estimating surface fluxes over the SGP site with remotely sensed data. Phys Chem Earth 25(2):167–172

Hanks RJ (1974) Model for predicting plant yield as influenced by water use. Agron J 66(5): 660–665

Jensen ME, Burman RD, Allen RG (1990) Evapotranspiration and irrigation water requirements, ASCE manual, vol 70. American Society of Civil Engineers, New York

Kulagina TB, Meshalkina JL, Florinsky IV (1995) The effect of topography on the distribution of landscape radiation temperature. Earth Obs Remote Sens 12:448–458

Kustas WP, Norman J (1999) Evaluation of soil and vegetation heat flux predictions using a simple two-source model with radiometric temperatures for partial canopy cover. Agric For Meteorol 94:13–29

Melesse AM, Nangia V (2005) Spatially distributed surface energy flux estimation using remotely-sensed data from agricultural fields. Hydrol Process 19(14):2653–2670

Melesse AM, Nangia V, Wang X, McClain M (2007) Wetland restoration response analysis using MODIS and groundwater data. Special issue: remote sensing of natural resources and the environment. Sensors 7:1916–1933

Morse A, Tasumi M, Allen RG, Kramber W (2000) Application of the SEBAL methodology for estimating consumptive use of water and streamflow depletion in the Bear River Basin of Idaho through remote sensing. Final report submitted to the Raytheon Systems Company, Earth Observation System Data and Information System Project, by Idaho Department of Water Resources and University of Idaho, 107 pp

Norman JM, Divakarla M, Goel NA (1995a) Algorithms for extracting information from remote thermal-IR observations of the earth's surface. Remote Sens Environ 51:157–168

Norman JM, Kustas WP, Humes KS (1995b) A two-source approach for estimating soil and vegetation energy fluxes from observations of directional radiometric surface temperature. Agric For Meteorol 77:263–293

Tanner B (1988) Use requirements for Bowen ratio and eddy correlation determination of evapotranspiration. In Proceedings of the specialty conference of the irrigation and drainage division. ASCE, Lincoln, NE, 19–21, July, 1988

Wang J, Bastiaanssen WGM, Ma Y, Pelgrum H (1998) Aggregation of land surface parameters in the oasis-desert systems of Northwest China. Hydrol Process 12:2133–2147

Chapter 12
Wetland Restoration Assessment Using Remote Sensing- and Surface Energy Budget-Based Evapotranspiration

Abstract Wetlands are one of the most important ecosystems with varied functions and structures. Humans have drained and altered the structure and functions of wetlands for various uses. Wetland restoration efforts require assessment of the level of ecohydrological restoration for the intended functions. Among the various indicators of success in wetland restoration, achieving the desired water level (hydrology) is the most important, faster to achieve, and easier to monitor than the establishment of hydric soils and wetland vegetation. Monitoring wetland hydrology using remote sensing-based evapotranspiration (ET) is a useful tool and approach since point measurements for understanding the temporal (before and after restoration) and spatial (impacted and restored) parts of the wetland are not effective. This chapter discusses the use of remote sensing and surface energy flux modeling approach to evaluate the state of wetland ET at two different wetland restoration sites: Glacial Ridge prairie restoration, northwestern Minnesota, and Kissimmee River basin, south Florida. Groundwater level and ET before and after the restoration is compared. Their spatial and temporal ET responses to the restoration activities were studied. Results show that the Landsat- and MODIS-based ET shows the level of spatiotemporal ET changes indicating an increase in ET values after the restoration.

Keywords Wetlands • Remote sensing • Evapotranspiration • Restoration • Glacial Ridge • Kissimmee River basin

12.1 Introduction

Wetlands are among the most valuable and productive ecosystems in the world. They improve water quality by absorbing and filtering out pollutants and sediments, store floodwaters, and slow down the force of flood and storm waters as they travel downstream. They offer habitat for wildlife and support biodiversity. The variety

W. Abtew and A. Melesse, *Evaporation and Evapotranspiration: Measurements and Estimations*, DOI 10.1007/978-94-007-4737-1_12,

of living organisms found in wetlands contributes to the health of our planet and our own lives. Wetlands also provide valuable open space and promote wonderful recreational opportunities.

Understanding hydrologic processes of wetlands is fundamental to their effective ecosystem restoration and creation (Mitsch and Gosselink 2000). According to the National Research Council (1995), one indicator of success of restored wetlands is the fulfillment of hydrology criteria such as flooding during the growing season. Research has shown that hydrologic processes such as hydroperiod, flow velocity, flow duration and variability, and evapotranspiration impact the ecosystem dynamics of wetlands (Cole and Brooks 2000; Gurnell et al. 2000; Price and Waddington 2000; Raghunathan et al. 2001; Janssen et al. 2004; Quinn and Hanna 2003).

Wetland restoration is designed to restore the functions and values of wetland ecosystems that have been altered or impacted through removal of vegetation, cropping, construction, filling, grading, and changes in water levels and drainage patterns. Activities and processes within and outside the wetland such as influx of sediments, fragmentation of a wetland from a contiguous wetland complex, loss of recharge area, or changes in local drainage patterns alter functions and structure of wetlands. The main goal of a wetland restoration is to restore the hydrology and vegetation back to their original condition or to ensure ecological integrity. The first step in wetland restoration is to restore the hydrology or water back to the wetland area. This involves halting the drainage of the wetland in the first place. Monitoring wetland hydrology recovery involves estimating surface and soil water budget, and ET changes at spatial scale.

The hydrologic functions that wetlands serve are often the basis upon which land management objectives are made for the entire watershed (Brooks et al. 1997). In most cases, wetlands exist where the groundwater table intersects, or is close to, the Earth's surface on a regular basis. As a result of the regular presence of water and dense vegetation, the highest ET rates within a watershed usually occur within wetlands. It is shown that a higher percentage of wetlands within a given watershed tends to decrease peak discharges during a given precipitation event indicating the value of wetlands in reducing peak flows through their regulation of flow.

Accurate quantification of ET with spatiotemporal domain has been a daunting task for hydrologists. This is mainly because ET is highly variable in space and time. Various methods exist for estimating ET. Traditional means such as the pan, Bowen ratio, eddy correlation, and aerodynamic techniques estimate ET at point locations. These methods are costly and time consuming and require elaborate and sensitive measurement equipment (Monteith and Unsworth 1990). A root zone soil water balance approach based on water budget is also a technique used to estimate ET as a residual variable. Quantifying each component of the soil water balance is less appealing in terms of time, labor, and money requirements. Relatively simpler point methods use lysimeter instrumentation (Brooks et al. 1997). While the traditional methods estimate ET at a point basis, recent methods have found success using remotely sensed imagery for estimates at various spatial scales (Tateishi and Ahn 1996; Mauser and Schadlich 1998).

In contrast to point measurements, remote sensing has the capacity to instantaneously acquire spectral signatures for large areas of the watershed in multiple electromagnetic (EM) wavebands and spatial scales. Data in multiple EM wavebands allows for the extraction of land cover, vegetation cover, emissivity, albedo, surface temperature, and energy flux information while data in regional scales allows for greater spatial coverage than possible with in situ methods.

Latent heat energy flux (evapotranspiration in energy units) is one of the most important components of the surface energy budget representing evaporation from the soil and transpiration from vegetation. Evapotranspiration in wetlands contributes to loss of water from the system. Assessing changes in wetland hydrology during wetland restoration requires knowledge of the magnitude of spatial ET at different stages of the restoration.

The section below discusses the application of remote sensing-based ET as an indicator of success in hydrologic wetland restoration projects. The study areas, methods, and data collected are discussed. The use of remote sensing-based data and surface energy flux modeling in spatial ET mapping is outlined. The ET before and after restoration is modeled and compared spatially along with other hydrologic variables such as groundwater level and biophysical changes using the satellite-based fractional vegetation cover. Two case studies outlining this application to the Glacial Ridge prairie restoration site in northwestern Minnesota and Kissimmee River restoration in south Florida are discussed below.

12.2 Case Studies

12.2.1 Glacial Ridge Prairie Restoration

Study Area Description: The Glacial Ridge prairie restoration project managed by The Nature Conservancy (TNC) is located northwest of Minnesota in Polk County (Fig. 12.1). The restoration covers 9,974 ha and hosts a great diversity of plant species including the threatened western prairie fringed orchid (TNC 2004). Other communities found at the preserve include wet and mesic tallgrass prairie and gravel prairie, willow thickets, mixed prairie, sedge meadow, aspen woodlands, and emergent marsh. The 30-year average mean annual precipitation of the site is 590 mm with annual average snow accumulation of 90 cm. The mean maximum and minimum monthly temperatures are 10 and −2.3°C, respectively with the highest temperatures occurring in July/August and the minimum in January/February. The geology of the site is characterized by glacial and postglacial deposits of Holocene and late Wisconsin age (Fullerton et al. 2004). According to TNC, in addition to its biological importance, the restoration of Glacial Ridge will help improve water quality for the city of Crookston and reduce flooding in the Red River Valley.

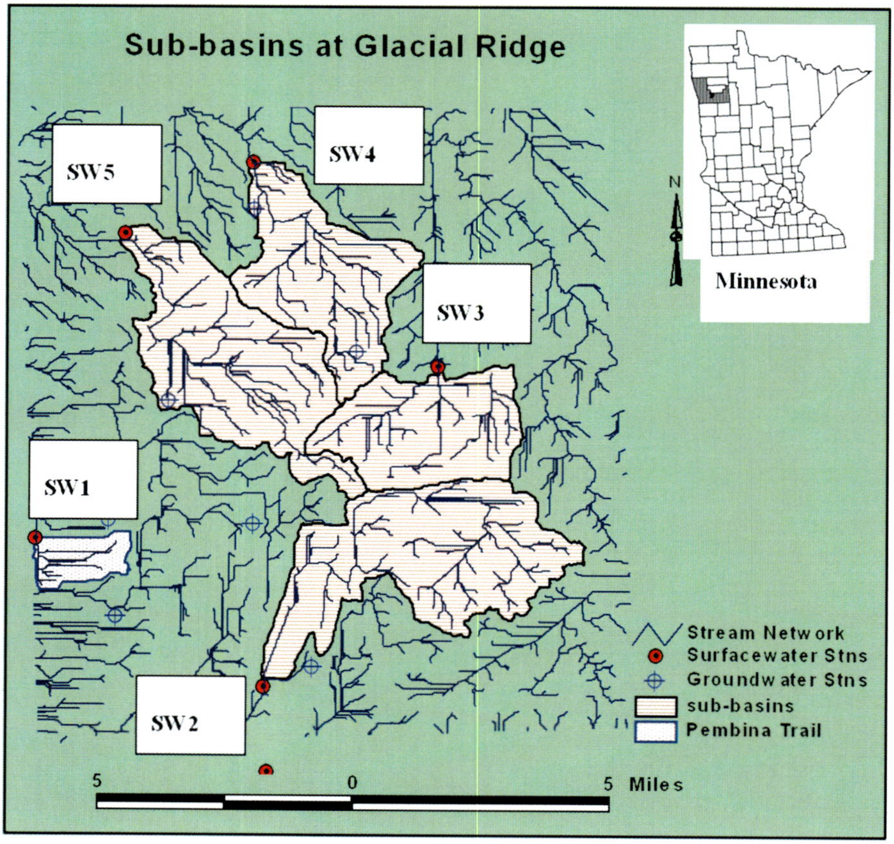

Fig. 12.1 Glacial Ridge prairie restoration site in Minnesota

In order to study the spatiotemporal variation in the hydrologic response of the study area to the restoration processes over the study years, five subbasins were delineated to represent different levels of impact or conditions of the wetlands (Fig. 12.1). These were subbasin 1 (SW1), subbasin 2 (SW2), subbasin 3 (SW3), subbasin 4 (SW4), and subbasin 5 (SW5). Pembina Trail (SW1) represents wetland with no impact. This subbasin served as the reference wetland for evaluating the hydrology of the other subbasins. Based on the data in 2002, SW2 and SW3 have undergone limited restoration in recent years. SW4 and SW5 are areas representing impacted wetlands with no restoration yet started.

Restoration Activities: TNC acquired the majority of the land in 2000 and 2001, and the restoration started in 2001. Some of the major restoration activities were divided into phases and implemented in the impacted areas (SW2 and SW3). The major restoration operations are land acquisition through purchase, planting native prairie species, burning exotic plant communities, and closures of farm and county ditches to restore the hydrology and raise the groundwater level. It is anticipated that

Table 12.1 List of Landsat images used in the Glacial Ridge prairie restoration study (Melesse et al. 2006)

Year	Date	Sensor	Path/row	Acq. time (UTM)	Local (CST) (UTM-6)
2000	June 5, 2000	L7 ETM+	30/27	17:08:37	11:08:37
	July 23, 2000	L7 ETM+	30/27	17:08:03	11:08:03
	August 24, 2000	L7 ETM+	30/27	17:07:45	11:07:45
2001	June 9, 2001	L5 TM	29/27	16:50:58	10:50:58
	July 10, 2001	L7 ETM+	30/27	17:06:08	11:06:08
	August 4, 2001	L7 ETM+	29/27	16:59:40	10:59:40
2002	June 4, 2002	L7 ETM+	29/27	16:59:11	10:59:11
	June 27, 2002	L7 ETM+	30/27	17:05:12	11:05:12
	July 29, 2002	L7 ETM+	30/27	17:05:00	11:05:00
2003	June 15, 2003	L5 TM	29/27	16:46:24	10:46:24
	July 24, 2003	L5 TM	30/27	16:53:20	10:53:20
	August 18, 2003	L5 TM	29/27	16:47:37	10:47:37

land acquired will convert from cropland to wetland reducing impact and bringing back the hydrology to ecologically favorable hydroperiod. Burning exotic plant species and planting native prairie plants will increase wetland biodiversity and plant community structure, which in turn ensures wetland health. Exotic plants tend to withdraw more water than the native species and thus lower groundwater level. The area has been drained for many years for the purpose of farming and on-site activities such as sand mining, installation of farmstead structures and roads. Closing the drainage ditches will raise the water level in the area maintaining the hydric soil property and also increases the ET of wetlands, thus setting the trajectory of the hydrology toward presettlement behavior.

Glacial Ridge Data Sets: The study used remotely sensed data (Landsat Thematic Mapper, TM, and Enhanced Thematic Mapper Plus, ETM+), topographic data (Digital Elevation Model, DEM) and weather data (solar radiation), wind speed, and air temperature.

Twelve Landsat TM and ETM + images from 2000 to 2003 (Table 12.1) for the months of June, July, and August were used to process the intermediate parameters (Normalized Difference Vegetation Index (NDVI)), radiometric surface temperature corrected using surface emissivity, albedo and surface and atmospheric emissivity from which surface energy flux components were estimated.

A 10-m DEM was used to delineate subbasins within the study area based on surface drainage. These subbasins represent areas of different stages of restoration activities and impact. The weather data were collected from on-site meteorological station managed by the United States Department of Agriculture Natural Resources and Conservation Service (USDA-NRCS) and the nearby weather station monitored by the North Dakota Agricultural Weather Network (NDAWN). The meteorological data, which were used as input to the surface energy flux model, include wind speed and air temperature. Hourly values are collected from the weather station at the study field. Only values at the time of the Landsat pass on the area were used. Figure 12.2 shows the average monthly solar radiation at the Glacial Ridge site.

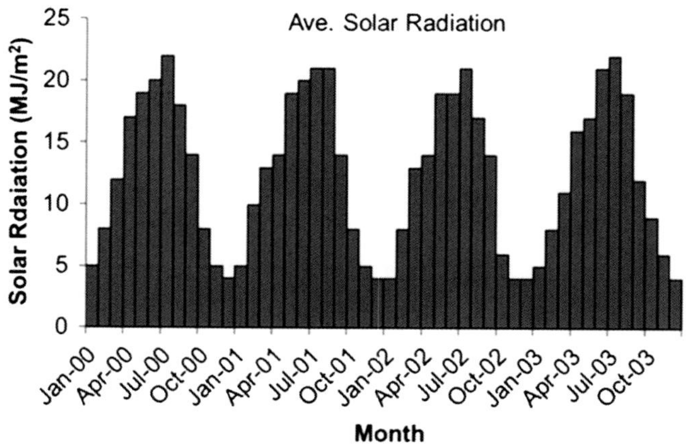

Fig. 12.2 Average monthly solar radiation at Glacial Ridge (2000–2003)

Fig. 12.3 Location of Kissimmee River basin in Florida

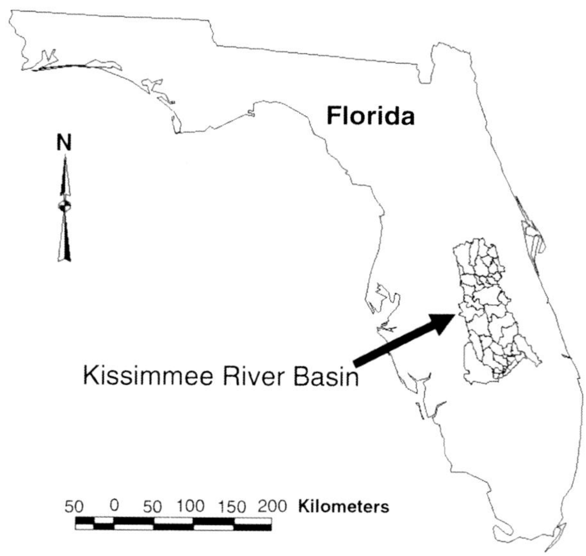

12.2.2 *Kissimmee River Restoration*

Study Area Description: The Kissimmee River basin is located north of Lake Okeechobee in south Florida (Fig. 12.3) and covers 7,680 km^2 and stretches from southern Orlando southward to Lake Okeechobee. The average annual rainfall of the subbasin is 120 cm. The wet season, with average rainfall of 76 cm, occurs from June through October, while the dry season, with average rainfall of 44 cm, occurs during the remaining months. On the average, the minimum and maximum air temperatures are 8.9 and 33.3°C occurring in January and July, respectively.

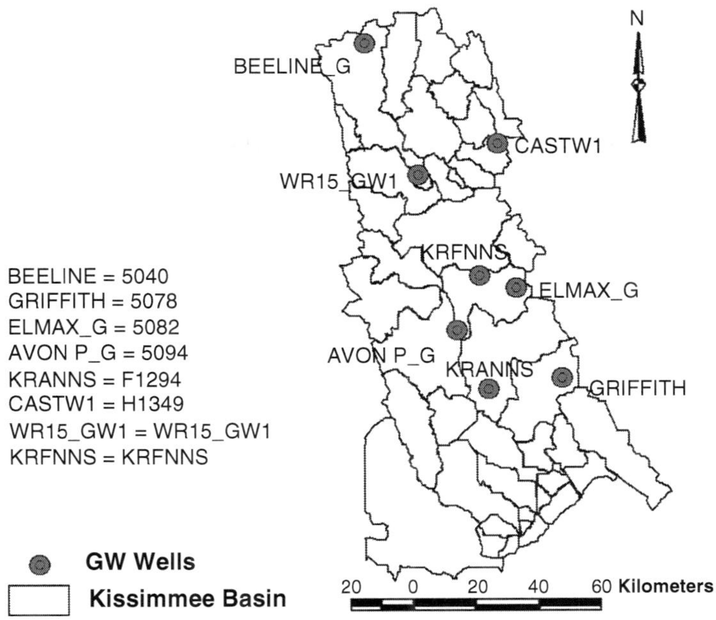

Fig. 12.4 Location of groundwater monitoring wells within the Kissimmee River basin

The average annual evapotranspiration of the Kissimmee basin is estimated to be 119.4 cm. The major land uses of the basin are wetlands, cropland, rangeland, and forested areas. The areal extent of these land-use classes has changed historically as a result of development and wetland drainage. The dominant wetland communities are broadleaf marsh, wet prairie, and wetland shrub.

Restoration Activities at Kissimmee River Basin: Successful restoration of the ecosystem requires ecohydrological integrity where the ecosystem is capable of supporting the biodiversity, value, and function of wetlands comparable to the natural level through the restoration of the hydrology and vegetation.

The Kissimmee River Restoration Project (KRRP) is working to reestablish the hydrologic conditions to recreate historical floodplains, wetland vegetation, and biodiversity and functionality through the removal of flood control canal, water control structures, and levees. The project expects the restoration of historical wetland ecosystems including a meandering river channel with a diversity of depths and wetland plant communities on the floodplain. The specific restoration activities undergoing since 1999 are rechannelization, revegetation, and land acquisition.

Data Sets: In order to analyze the effect of restoration on the groundwater levels, monthly water level was analyzed from eight selected monitoring wells from 2000 to 2004. Wells were selected to represent the different parts of the basin. The location of the selected monitoring wells in this study is shown in Fig.12.4.

The study considered at assessing the spatiotemporal changes of vegetation cover and latent heat flux (evapotranspiration in energy units) of the basin. Remotely sensed images from the Moderate Resolution Imaging Spectroradiometer (MODIS) aboard the Terra sensor was used in this study. Images for the months of April, September, and December from 2000 to 2004 were acquired and processed. Daily surface temperature, normalized difference vegetation index (NDVI), and albedo were also acquired from the Land Processes Distributed Active Archive Center (LP DAAC) and used in the surface energy balance computation. Micrometeorological data collected from the National Climatic Data Center (NCDC) include air temperature and wind speed.

12.3 Methodology

12.3.1 *Satellite Image Preprocessing*

Images were georeferenced and calibrated to ground reflectance using Eq. 12.1:

$$R_x = \frac{\pi d^2(\text{Gain}_x{}^*(d_{\text{e}-\text{s}}N_x - \text{min}_x))}{\text{Esun}^* \sin\theta} \tag{12.1}$$

where R is the ground reflectance for each band (x), $d_{\text{e-s}}$ is the Earth–Sun distance in astronomical unit for the image date, Gain is the solar spectral irradiance for each band, DN represents digital number in the raw image, min is the lower DN in the specific band (Chavez 1996), Esun is the mean solar exoatmospheric irradiance, and θ is the sun elevation angle (Huang et al. 2002).

12.3.2 *Evapotranspiration Mapping*

Remote sensing-based ET estimations using the surface energy budget equation are proving to be one of the most recently accepted techniques for areal ET estimation (Morse et al. 2000). Surface Energy Balance Algorithms for Land (SEBAL) is one of such models utilizing Landsat images and images from others sensors with a thermal infrared band to solve Eq. 10.1 in Chap. 10 and hence generate areal maps of ET (Bastiaanssen et al. 1998a, b; Morse et al. 2000).

SEBAL requires weather data such as solar radiation, wind speed, precipitation, air temperature, and relative humidity in addition to satellite imagery with visible, near infrared, and thermal bands. SEBAL uses the model routine of ERDAS Imagine in order to solve the different components of the energy budget equations. Figure 12.5 shows the evapotranspiration computation flowchart using the surface energy flux balance approach. Using Eqs. 10.25 and 10.26 in Chap. 10, fractional vegetation cover (FVC) is mapped, and comparisons are made for the Kissimmee River restoration case study.

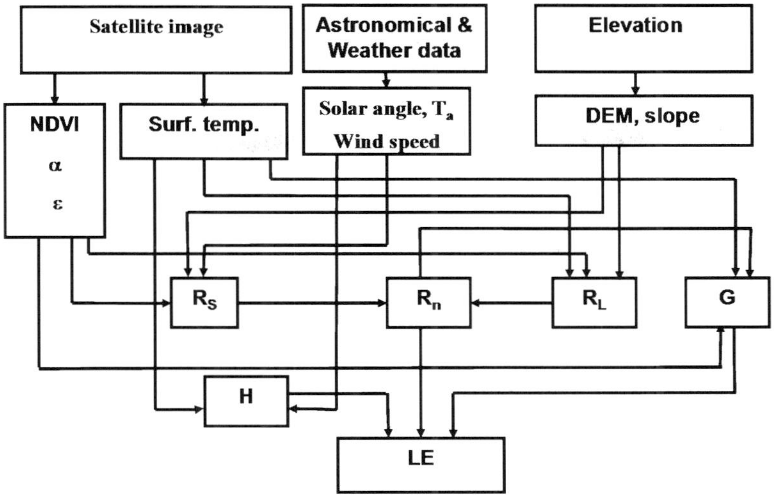

Fig. 12.5 Evapotranspiration mapping flowchart (Melesse et al. 2006)

12.4 Results and Discussion

12.4.1 Glacial Ridge

12.4.1.1 Verification

The 24-h ET (ET_{24}) image was checked for agreement with the weather station24-h reference ET (ET_{r24}), computed from observed on-site weather station data at the time of image capture. Table 12.2 shows the ET and surface temperature calibration results.

An average error of prediction using the SEBAL approach was −4.3%. Table 12.2 shows reliable agreement between the ET_{r24} and the ET_{24} estimated by SEBAL for most of the ETM + images, the highest discrepancy being 13.7% with the August 4, 2001, image. For the July 2001 image, some cloud cover occupied the line of sight between the sensor and the weather station and thus gave erroneous results. The TM images gave the highest discrepancies during 2003 (>20%). It is not clear why such disagreements existed in 2003, while the 2000–2002 images produced reasonable results (<7.7%), although slight striping effects were observed for the September 2003 image, possibly indicating sensor radiometric inconsistencies.

Table 12.2 Observed and SEBAL 24-h ET and surface temperature comparisons at Glacial Ridge (Oberg and Melesse 2006)

Date	Path/row	Sensor	ET_{r24} reference (mm)	ET_{24} SEBAL (mm)	ET_{24} error (%)	T_s observed[a] (°C)	T_s SEBAL (°C)	T_s error (%)
June 5, 2000	30/27	ETM+	7.4	6.9	−6.1	19.8	23.8	−1.37
July 23, 2000	30/27	ETM+	6.6	7.1	8.3	24.8	27.8	−1.01
August 24, 2000	30/27	ETM+	5.1	5.1	−0.4	22.8	25.8	−1.01
June 9, 2001	29/27	TM	8.7	8.5	−2.7	24.8	25.8	−0.34
July 10, 2001[b]	30/27	ETM+	6.7	6.1[b]	−8.5	22.8	29.8	−2.36
August 4, 2001	29/27	ETM+	6.0	5.1	−13.7	29.8	28.8	0.33
June 4, 2002	29/27	ETM+	5.0	4.7	−6.6	16.8	24.8	−2.76
June 27, 2002	30/27	ETM+	6.2	5.9	−4.7	26.8	27.8	−0.33
July 29, 2002	30/27	ETM+	6.1	5.6	−7.9	23.8	25.8	−0.67
June 15, 2003	29/27	TM	7.5	7.5	0.1	23.8	20.8	1.01
July 24, 2003	30/27	TM	7.4	5.8	−22.3	23.8	25.8	−0.67
August 18, 2003	29/27	TM	6.8	5.3	−22.2	29.8	30.8	−0.33

[a]Near surface temperature at weather station

[b]Cloud cover observed. Nearest representative pixel selected

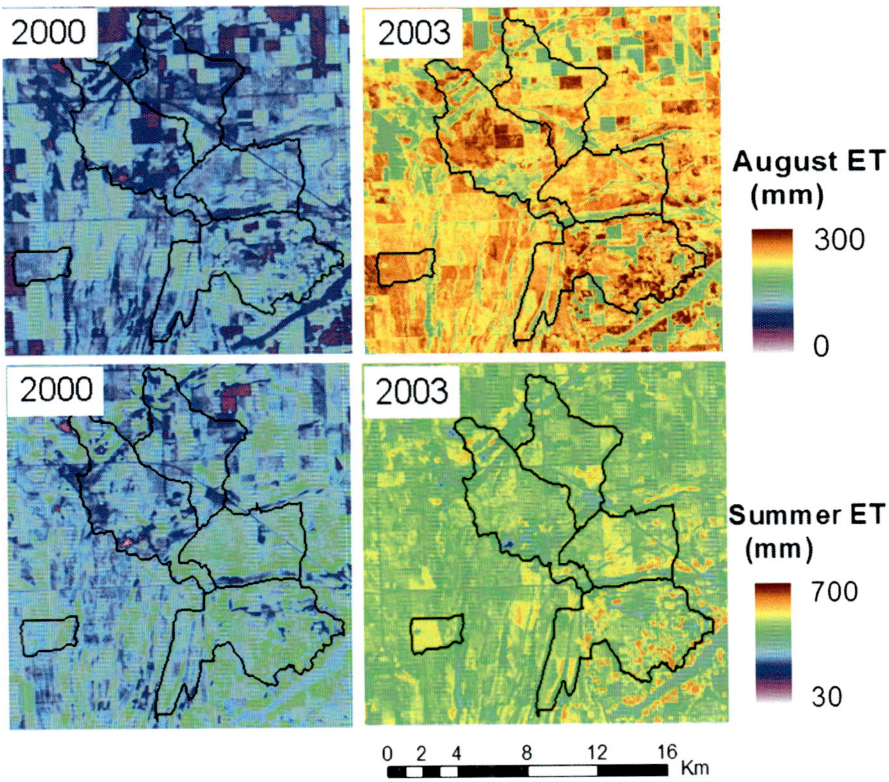

Fig. 12.6 Spatial ET maps for the month of August and summer (June, July, and August) 2000 and 2003 at Glacial Ridge (Melesse et al. 2006)

12.4.1.2 Spatial Evapotranspiration Changes

In order to assess the effect of wetland restoration on ET changes, understanding the variation of the monthly solar radiation from 2000 to 2003 at the study site is necessary (Fig. 12.2). The average solar radiation (June–August) was similar among the years ranging from 19 to 20.7 MJ m^{-2}.

Spatial ET (monthly and seasonal) was mapped for 2000 and 2003 for August and summer (Fig. 12.6). It is shown that ET has both spatial and temporal changes at the study site reflecting the effect of the restoration on the ET response. SW1 has shown less variation in the temporal ET than the other subbasins over the study years. Similarly, the mean monthly ET of SW1 was higher than that of SW4 and SW5 (Table 12.3). The mean monthly and seasonal ET for SW2 and SW3 was also higher than that of the SW4 and SW5.

The restoration of the wetlands at Glacial Ridge started in 2001. Spatial ET of 2000 was used as a reference for assessing ET changes for 2001–2003. As shown in Fig. 12.6, both the August and seasonal ET were higher in 2003 than 2000 due to

Table 12.3 Average seasonal (June–August) evapotranspiration (mm) at Glacial Ridge (Melesse et al. 2006)

Year	SW1	SW2	SW3	SW4	SW5	Ave.
2000	320	340	340	290	300	320
2001	360	360	360	340	320	350
2002	450	460	440	420	400	430
2003	490	480	480	460	450	470
Ave.	410	410	410	380	370	

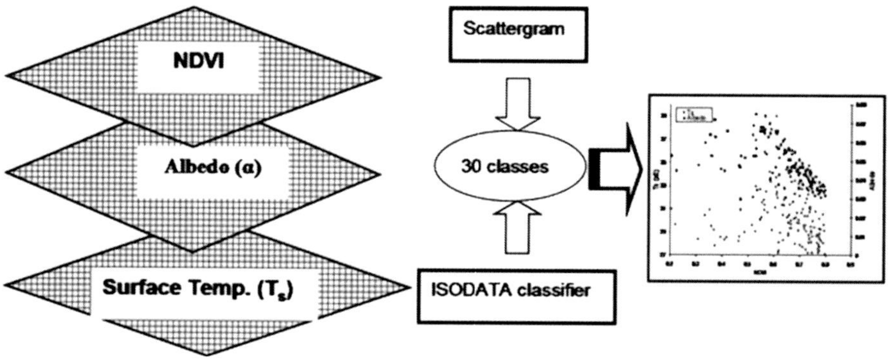

Fig. 12.7 Flowchart for the scattergram development (Melesse et al. 2007)

restoration activities. The average annual ET increases for the five subbasins were in the range of 9% (2000–2001 and 2002–2003) and 25% (2001–2002). Between 2000 and 2003, ET increased by at an average of 50% across the study areas.

12.4.2 Kissimmee River Basin

12.4.2.1 NDVI–T_S–Albedo Relationship

From MODIS data, monthly values of NDVI, T_S, and albedo were generated for the months of April, September, and December from 2000 to 2004. The selection of the months was designed to represent the different times of the year. Using these values as layers, unsupervised classification was run using the iterative self-organizing data analysis (ISODATA) algorithm (ERDAS 1999). This classification yielded 30 classes for each month. Combining the resulting land-cover classes from each run (3 months × 5 years) gave a scattergram of NDVI–T_S–albedo (Fig. 12.7). Figure 12.8 shows the scattergram. It is shown that surface temperature and albedo have a negative relationship with the level of green vegetation especially for NDVI >0.5 with R^2 value of 0.61 and 0.15, respectively (Fig. 12.8). Higher latent heat losses from the vegetated surface lead to a cooler surface and lower surface temperature in vegetated areas than bare ground. This relationship is not clearly defined in less vegetated surfaces (water bodies and bare ground) as shown in

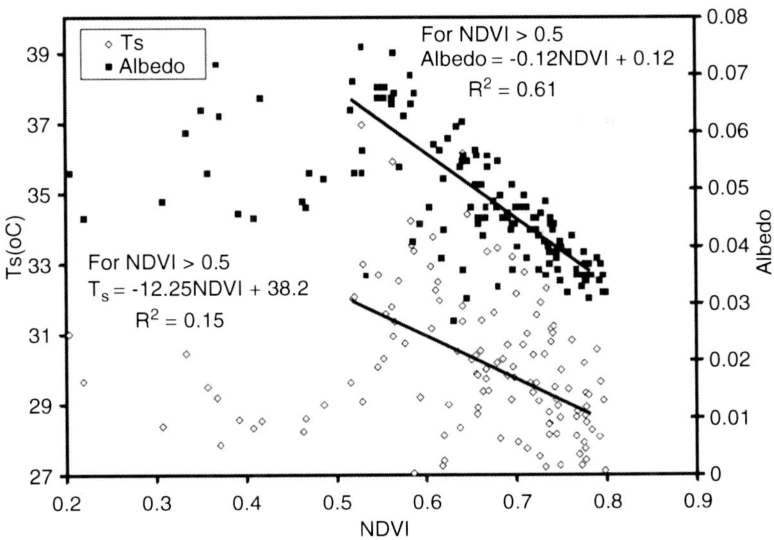

Fig. 12.8 Scattergram of albedo, NDVI, and surface temperature (Melesse et al. 2007)

the left-hand side of the graph (Fig. 12.8). Similarly, albedo has also a negative correlation with NDVI for the highly vegetated portion of the basin. For less vegetated surfaces, this relationship is not distinct (Fig. 12.8).

12.4.2.2 Fractional Vegetation Cover Changes

Fractional vegetation cover (FVC) for the month of April from 2000, 2002, and 2004 was generated, and comparisons were made (Fig. 12.9). It is shown that FVC has shown changes along the river, especially in the middle portion of the watershed. Although changes are not significant (mean April FVC of 0.15, 0.16, and 0.17 for 2000, 2002, and 2004, respectively), the trend is an indicator of some response of the vegetation along the river to the restoration work (Table 12.4). The actual changes in the FVC will require field sampling and close observation. This study does not identify the type of vegetation and if this response is a desirable one. Table 12.4 shows statistics of the FVC for the period of the study for the area along the river as shown in Fig. 12.9.

12.4.2.3 Latent Heat Flux Dynamics

Latent heat grids were generated from MODIS imagery for the month of April (2000, 2002, and 2004). Figure 12.10 show maps of latent heat in watts per square meter. As it is depicted in Fig. 12.10, latent heat values were higher in 2002 and

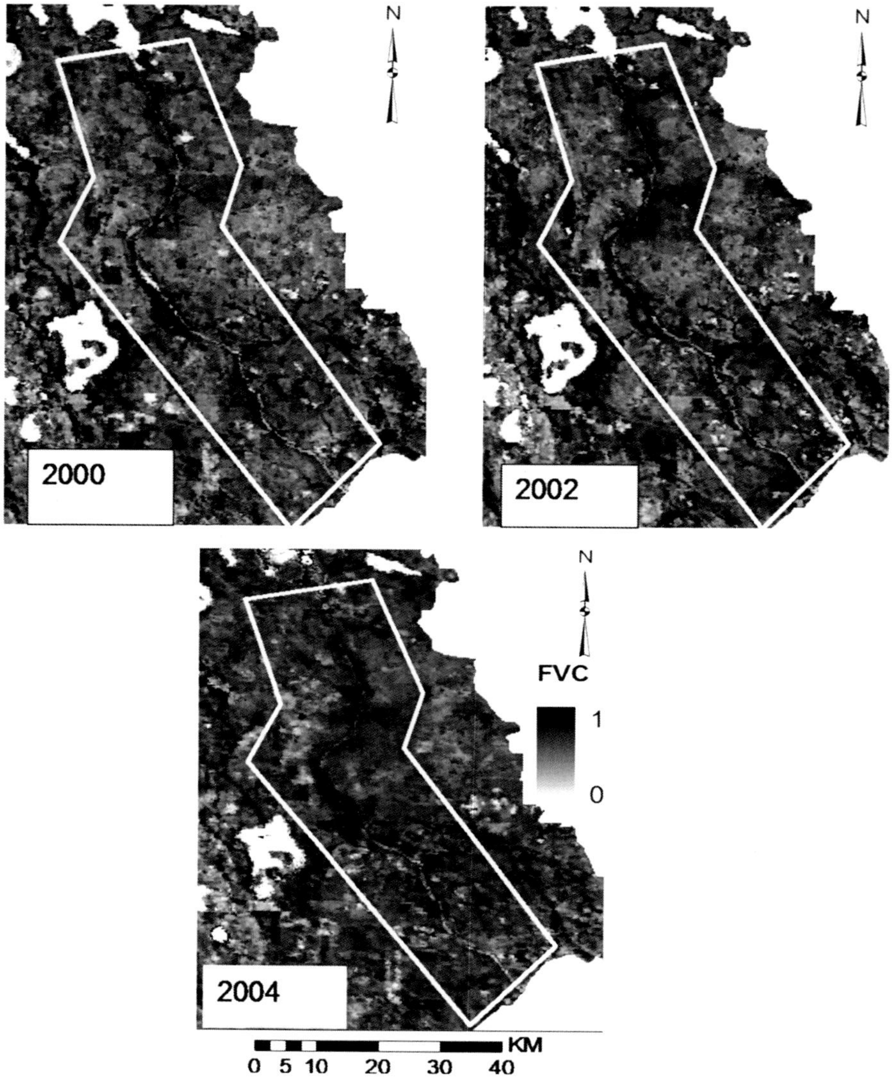

Fig. 12.9 MODIS-based fractional vegetation cover (FVC) of the area of interest within the Kissimmee River basin for 2000, 2002, and 2004 (Melesse et al. 2007)

2004 than 2000 on areas along the rivers (Table 12.4). The average April LE for 2000, 2002, and 2004 were 128, 135, and 139 W m^{-2}, respectively. The removal of flood control structures and rechannelization of the river to its natural course will increase the floodplain area and in turn lead to higher latent heat flux. It is shown that higher latent heat flux along the river can be attributed to the increased flood plain areas and vegetation cover. The rainfall volume for the month of April (2000, 2002, and 2004) was 40, 10, and 35 mm, respectively.

Table 12.4 Statistics of average FVC and monthly LE values for the month of April for 2000, 2002, and 2004 for the area of interest along the Kissimmee River, south Florida (Melesse et al. 2007)

Year	Min	Max	Mean	Std. dev.
Average April FVC				
2000	0	0.97	0.15	0.20
2002	0	0.81	0.16	0.21
2004	0	0.90	0.17	0.21
April LE (W m^{-2})				
2000	0	594	128	189
2002	0	597	135	166
2004	0	664	139	169

12.4.2.4 Groundwater Data

Groundwater level is an indicator of the response of wetlands to restoration. Change in hydrology of wetlands with a shallow groundwater table is used as one of the measure of success of the restoration activity. Eight groundwater monitoring wells were selected to represent the different locations in the basin. Monthly groundwater level from these wells was used, and comparisons were made for the period of study. Taking into account the volume of rainfall for each year, it is shown that wells along the rivers have shown a shallower groundwater table between 2001 and 2003 (Table 12.5) compared to other years. It was also shown that analysis of groundwater level data (2000–2004) from eight monitoring wells showed that the average monthly level of groundwater was increased by 20 and 34 cm between 2000 and 2004 and 2000 and 2003, respectively (Table 12.5).

12.5 Summary

12.5.1 Glacial Ridge

The effect of the restoration on the hydrologic regime changes and hence on the spatial ET was studied using a surface energy budget technique from a remote sensing perspective. Comparative study of the different subbasins of the restoration site for their hydrologic response shows that recent restoration activities increased the water tables and hence the spatial ET. Over the study period, ET increased nearly by 50% over the study area with an average annual increase of 14%. Such an approach of assessing the ecohydrological restoration from a remote sensing perspective is useful and applicable. The study only considered ET changes as criteria for evaluating the changes that occurred. Incorporating ground- and surface water data in the watershed's spatial surface and soil water budget will help in understanding the changes fully.

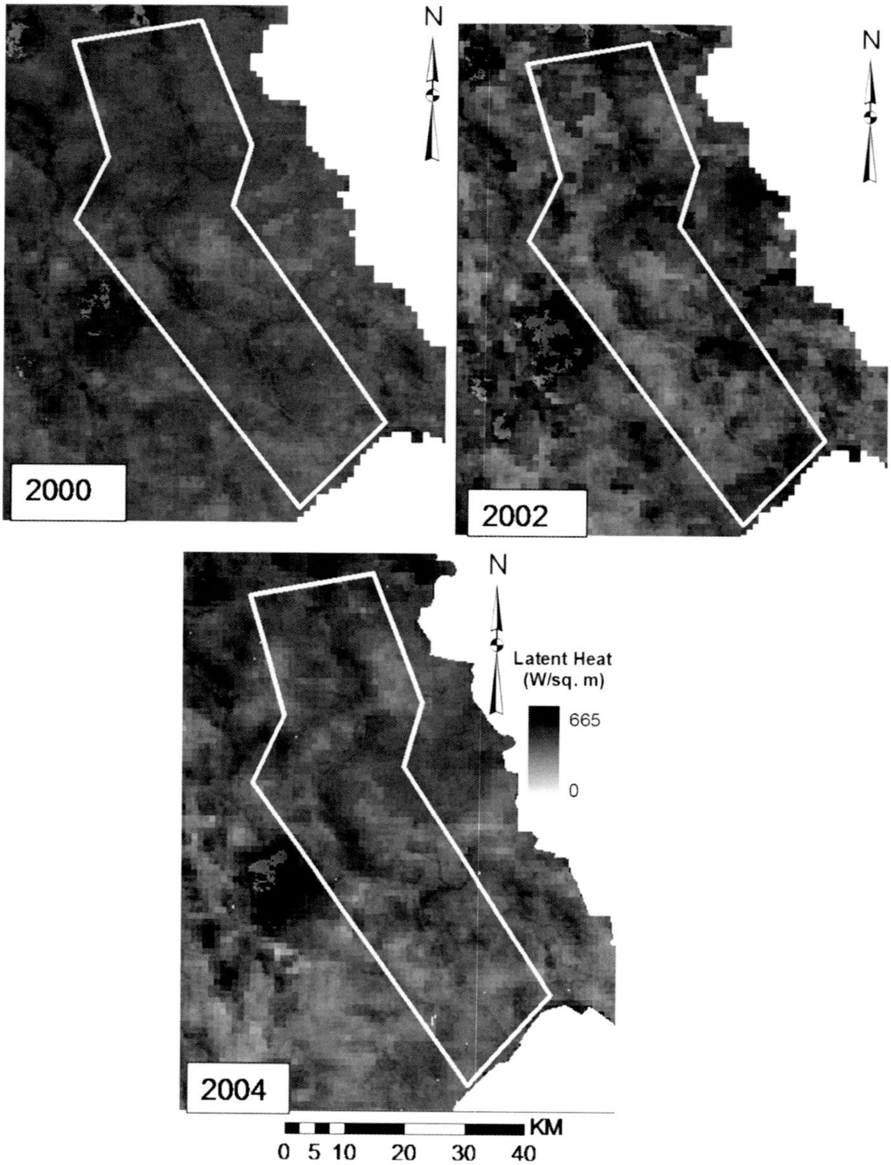

Fig. 12.10 MODIS-based monthly latent heat flux (W m^{-2}) of the area of interest within the Kissimmee River basin for 2000, 2002, and 2004 (Melesse et al. 2007)

12.5.2 Kissimmee River Basin

Response of the Kissimmee basin's hydrology and vegetation to the recent restoration was evaluated using data from MODIS-based FVC, spatial latent heat flux,

Table 12.5 Average monthly groundwater levels (m amsl) for selected wells in the Kissimmee River basin (Melesse et al. 2007)

Year	5040	5078	5082	5094	F1294	KRFNNS	H1349	WR15	Ave. mon. GW level (m amsl)	Ave. mon. rainfall (mm)
2000	27.09	21.72	21.49	41.85	12.33	15.11	22.72	19.80	22.76	23.98
2001	26.80	21.83	17.90	41.92	11.83	15.01	23.02	19.92	22.28	28.87
2002	27.31	21.91	21.61	42.16	11.89	15.30	23.03	20.26	22.93	24.83
2003	27.87	21.90	21.72	42.14	12.28	15.40	23.04	20.47	23.10	30.79
2004	27.73	21.82	21.57	42.00	12.28	15.29	23.00	20.17	22.96	34.73
Ave.	27.36	21.84	20.86	42.02	12.12	15.22	22.96	20.12	22.81	28.64

and groundwater records. The NDVI–T_S–albedo relationship was also analyzed for the 2000–2004 period. Using NDVI, T_S, and albedo values for the month of April, unsupervised classification was conducted and a scattergram was generated. Results show that for the highly vegetated portion of the graph, a negative correlation between NVDI–T_S and NDVI–albedo was observed. It was also indicated that for the less vegetated (lower NDVI) part, the NDVI–T_S–albedo relationship was not clearly defined.

The fractional vegetation cover was increased for 2002 and 2004 compared to 2000 for areas along the Kissimmee River indicating response to the floodplain restoration. The spatial latent heat flux, which is evapotranspiration in energy units, has also shown an increase in 2002 and 2004 compared to 2000, which can be attributed to large areas of vegetated surface. This change was mainly seen along the river where most of the restoration work is occurring and changes in the hydrology are expected.

The groundwater level records from selected monitoring wells were also used to compare spatiotemporal variations in the groundwater levels. Analysis of groundwater level data (2000–2004) from eight monitoring wells showed that the average monthly level of groundwater was increased by 20 and 34 cm between 2000 and 2004 and 2000 and 2003, respectively. Taking into account the amount of rainfall, this observation is valid and reasonable. Understanding the complete ecohydrological response of the basin due to the restoration work will require collection and analysis of vegetation cover at finer scales than reported in this study.

References

Bastiaanssen WGM, Menenti M, Feddes RA, Holtslag AAM (1998a) The surface energy balance algorithm for land (SEBAL). Part 1: formulation. J Hydrol 212–213:198–212

Bastiaanssen WGM, Pelgrum H, Wang J, Ma Y, Moreno J, Roerink GJ, van der Wal T (1998b) The surface energy balance algorithm for land (SEBAL). Part 2: validation. J Hydrol 212–213:213–229

Brooks KN, Ffolliott PF, Gregersen HM, DeBano HF (1997) Hydrology and the management of watersheds, 2nd edn. Iowa State University Press, Ames

Chavez PS (1996) Image-based atmospheric correction – revised and improved. Photogramm Eng Remote Sens 62(9):1025–1036

Cole CA, Brooks RP (2000) A comparison of the hydrologic characteristics of natural and created mainstream floodplain wetlands in Pennsylvania. Ecol Eng 14:221–231

Earth Resources Data Analysis System (ERDAS) (1999) ERDAS field guide. ERDAS Inc., Atlanta

Fullerton D, Bush C, Pennell J (2004) Map of surficial deposits and materials in the eastern and central United States (East of 102° West Longitude). U.S. Geological Survey Geologic Investigations Series I-2789. U.S. Geological Survey, Denver, CO

Gurnell AM, Hupp CR, Gregory SV (2000) Linking hydrology and ecology. Hydrol Process 14:2813–2815

Huang, C, Zhang Z, Yang L, Wylie B, Homer C (2002) MRLC 2000 image preprocessing procedure. USGS white paper. EROS Data Center, Sioux Falls

Janssen R, Goosen H, Verhoeren ML, Verhoeren JTA, Omtzigt AQA, Maltby E (2004) Decision support for integrated wetland management. Environ Model Softw 20(2):215–229. doi:10.1016/j.envsoft. 2003.12.020

Mauser W, Schadlich S (1998) Modeling the spatial distribution of evapotranspiration on different scales using remote sensing data. J Hydrol 212–213:250–267

Melesse AM, Oberg J, Beeri O, Nangia V, Baumgartner D (2006) Spatiotemporal dynamics of evapotranspiration and vegetation at the Glacial Ridge Prairie restoration. Hydrol Process 20(7):1451–1464

Melesse AM, Nangia V, Wang X, McClain M (2007) Wetland restoration response analysis using MODIS and groundwater data special issue: remote sensing of natural resources and the environment. Sensors 7:1916–1933

Mitsch WJ, Gosselink JG (2000) Wetlands. Wiley, New York

Monteith JL, Unsworth MH (1990) Principles of environmental physics, 2nd edn. Butterworth–Heinemann, Woburn

Morse A, Tasumi M, Allen RG, Kramber W (2000) Application of the SEBAL methodology for estimating consumptive use of water and stream flow depletion in the Bear River Basin of Idaho through remote sensing. Final report submitted to the Raytheon Systems Company, Earth Observation System Data and Information System Project, by Idaho Department of Water Resources and University of Idaho

National Research Council (1995) Wetlands characteristics and boundaries. National Academy Press, Washington, DC

Oberg J, Melesse AM (2006) Evapotranspiration dynamics at an ecohydrological restoration site: an energy balance and remote sensing approach. J Am Water Resour Assoc 42(3):565–582

Price JS, Waddington JM (2000) Advances in Canadian wetland hydrology and biogeochemistry. Hydrol Process 14:1579–1589

Quinn NWT, Hanna WM (2003) A decision support system for adaptive real-time management of seasonal wetlands in California. Environ Model Softw 18:503–511

Raghunathan R, Slawecki T, Chen ZQ, Dilks DW, Bierman VJ, Wade S (2001) Exploring the dynamics and fate of total phosphorus in the Florida Everglades using a calibrated mass balance model. Ecol Model 142:247–259

Tateishi R, Ahn CH (1996) Mapping evapotranspiration and water balance for global land surfaces. J Photogramm Remote Sens 51:209–215

The Nature Conservancy (TNC) (2004) http://nature.org/wherewework/northamerica/states/minnesota/preserves/art6943.html. Accessed 25 Mar 2004

Chapter 13
Climate Change and Evapotranspiration

Abstract Climate change has been acknowledged as one of the greatest challenges for humanity. Although, there may be differences in opinions as to the cause of change, it is generally accepted that climate change is happening. There is sufficient data showing sea level rise and temperature rise and associated ecological changes. Climate change impacts on rainfall and evapotranspiration have not been conclusively determined. Decrease in rainfall and increase in temperature will result in increase in evapotranspiration. Global circulation models' (GCMs) applications have shown spatially varying diverse trends for evapotranspiration. It is essential to put forth the effort to know the impact of climate change on evapotranspiration and use the information for developing adaptations in water use and water management.

Keywords Climate change • Evapotranspiration • Global circulation models • South Florida • Great River Basin Jamaica

13.1 Introduction

Climate change, especially global warming, is expected to impact ecosystems negatively. Expected outcomes are rising sea levels and rising temperature in most regions. Impact on precipitation, evapotranspiration, and runoff is widely believed to vary by region. The worst-case scenario for water supply is decreasing rainfall and increasing evapotranspiration. Evapotranspiration increases with increasing temperature, increasing radiation, decreasing humidity, and increasing wind speed. Decreasing rainfall contributes to increasing evapotranspiration through increase in clear skies, increase in temperature, and lower humidity. Regional evaluation of climate change impact is necessary to evaluate impacts on evapotranspiration, precipitation, and runoff.

W. Abtew and A. Melesse, *Evaporation and Evapotranspiration: Measurements and Estimations*, DOI 10.1007/978-94-007-4737-1_13,

13.2 Climate Change and Evapotranspiration

The United Nations Intergovernmental Panel 2007 report on climate change clearly shows warming trends. Forms of mitigation of the change and adaptation to the change have started developing in many places. Considerations are being given to climate change in infrastructural plans. Evapotranspiration is a main component of the global water and energy cycle. A change in climate and weather parameters will result in a significant change in evapotranspiration. Global warming can directly affect evapotranspiration through increase in radiation, rise in temperature, and increase in water vapor deficit. The results of a global climate model (GCM) simulations for three Alpine river basins for summer temperature increase of 3–4°C was found to increase potential evapotranspiration by 20 (Calanca et al. 2006). The predicted increase in solar radiation was 5, and 10–20% precipitation decrease was anticipated. Based on 317 weather station data analysis in China, it was found that evaporation has increased since 1980 with global warming (Cong et al. 2008). A report by the Union of Concerned Scientists and the Ecological Society of America predicts based on climate models that the Gulf Coast of the United States temperature will increase between 3 and 7°F for summer highs and 5 to 10°F for winters (Twilley et al. 2001). Higher temperature will increase evapotranspiration. A concern on the impact of water vapor contribution from evapotranspiration on global warming and the need to grow water efficient crops is presented by Azam and Farooq (2005). A study on the effect of global warming on evapotranspiration of alfalfa production in California applied a global circulation model and weather simulation model. The results indicated a prediction of daily mean maximum temperature increase of 4.3°C and statewide mean daily evapotranspiration increase of 0.59 mm (Zhang et al. 1996).

Based on 15-model mean changes, the Intergovernmental Panel on Climate Change reported an estimated 15% increase in annual evaporation for south Florida for the period 2080–2099 compared to 1980–1999 (Bates et al. 2008). Application of the Canadian Regional Climate Model to evaluate hydrologic impacts of climate change produced results of as high as a 20 cm (8 in.) increase in reference evapotranspiration with mostly 7–13 cm (3–5 in.) general increase over the whole state of Florida (Fig. 13.1, Obeysekera et al. 2011). Reference evapotranspiration was computed based on the Penman–Monteith equation.

The effects of global warming on south Florida evapotranspiration will be of a similar trend, increasing. Evaporation and evapotranspiration have been shown to have a direct relationship with solar radiation and air temperature (Abtew 1996). Increase in CO_2, solar radiation, and temperature will result in an increase in crop and vegetation productivity and water use. The increase in evapotranspiration should be of sufficient concern to warrant studies to estimate the increases and incorporate in water management strategies.

The Special Report on Emissions Scenarios (SRES) are grouped into four scenario families (A1, A2, B1, and B2) that explore alternative development pathways, covering a wide range of demographic, economic, and technological

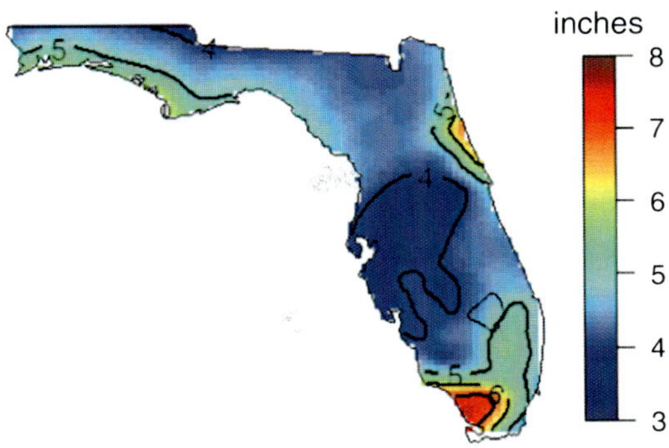

Fig. 13.1 Change in reference evapotranspiration in Florida by 2050 (Obeysekera et al. 2011; provided by South Florida Water Management District); 1 in. = 2.54 cm

driving forces and resulting GHG emissions (IPCC 2007). Based on these scenarios, various analyses for the watershed scale simulation of potential impacts of climate change on the water budget components have been done in different parts of the world. Although the reliability of the various downscaling techniques from large-scale GCM products to the watershed scale is different, it has been an acceptable practice to potentially understand the river basin impacts.

Hydrological impact studies rely on GCM outputs for watershed scale assessment of potential hydrologic alterations emanating from climate change-related variations in precipitation and air temperature. This will necessitate the downscaling of the large-scale GCP outputs to watershed scale. The downscaled outputs are then used as inputs to hydrological models for predicting the changes in stream flow, groundwater availability, evapotranspiration, and other water budget components.

Based on application of three global circulation models (Canadian Center for Climate Modeling and Analysis (CCCMA), Canada, Geophysical Fluid Dynamics Laboratory (GFDL), USA and Max-Planck-Institute for Meteorology (MPI-M), Germany), for three scenarios (A1B, A2, and B1) (IPCC 2007). Air temperature and rainfall were downscaled to the Great River region of Jamaica in an effort to assess the impacts of climate change on watershed scale hydrology. Using the Soil Water Assessment Tool (SWAT) (Arnold et al. 1998) along with stream flow, groundwater, and others, potential ET was predicted for the Great River basin, Jamaica (Melesse et al. 2011). Based on the modeling result, an overall average of 15 mm $month^{-1}$ increase in potential evapotranspiration is projected for the period of 2080–2100. Figure 13.2 shows the average watershed scale-predicated air temperature increase from the base period, 1980–2000. The results project an overall average of 2.36°C increase in air temperature (Fig. 13.2). Figure 13.3 depicts monthly mean potential increase of potential ET for each scenario averaged from outputs of the three models.

There are studies that report plant transpiration increasing with temperature but a significant increase in CO_2 reduces the increase in plant transpiration attributed to an increase in temperature (California Department of Water Resources 2006; Hatfield et al. 2008). Accordingly, open water and soil evaporation does not decrease with an increase in CO_2 while increasing with increase in temperature. A USDA study on effects of climate change on agriculture evaluated the impact of CO_2 on crop

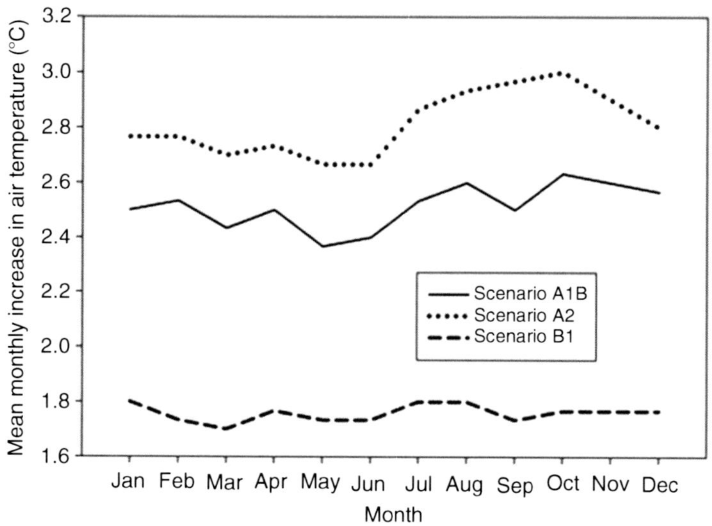

Fig. 13.2 Expected air temperature increase in the Great River region of Jamaica (2080–2100)

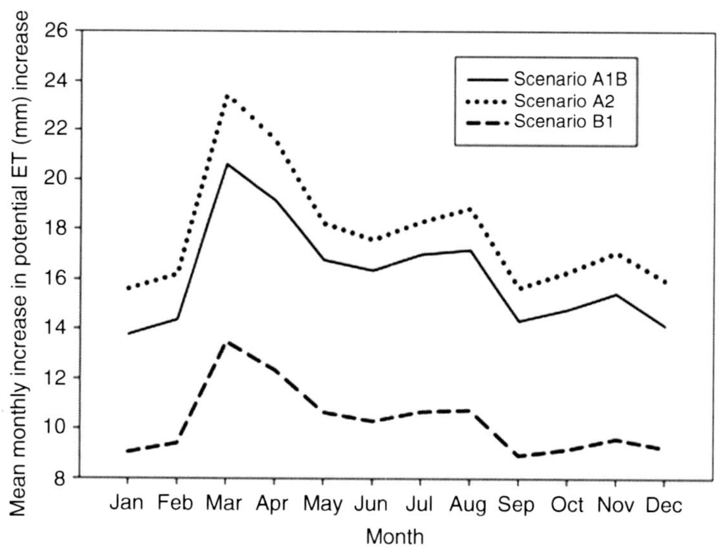

Fig. 13.3 Expected potential evapotranspiration increase in the Great River region of Jamaica (2080–2100)

evapotranspiration in the United States without considering the change in temperature. It concluded that with ample nitrogen and limited water, evapotranspiration will stay the same for both C3 and C4 plants. But, with ample nitrogen and ample water, reduction in evapotranspiration is projected at 550 ppm CO_2 concentrations

(Hatfield et al. 2008). Contrary to these projections, based on modeling impacts of climate change on water resources of Finland, evapotranspiration increases of 6, 13, and 23% are reported for 2020, 2050, and 2010 (Vehviläinen and Huttunen 1997). Sensitivity of evapotranspiration to global warming in a study for the arid region of Rajasthan in India concludes that a marginal increase in ET on such climatic area will have significant impact (Goyal 2004). A study of climate change, evaporation, and evapotranspiration using six global climate models projected the likelihood of increased potential evapotranspiration over India (Chattopadhyay and Hulme 1997). A global study of the Palmer drought index relationship to soil moisture and effects of surface warming concluded that as anthropogenic global warming progresses, risk of droughts will increase due to increased temperature and increased drying (Dai et al. 2004).

According to Jung et al. (2010), more than half of the solar energy absorbed by land surfaces is used for evaporation. They project that climate change will alter evapotranspiration through changes in the hydrologic cycle. Analyzing global meteorological monitoring network data, and remote sensing data and applying modeling, they concluded that evapotranspiration increased by 7 mm year^{-1} between 1982 and 1997. Also, they concluded that the decline in evaporation from 1997 to 2008 was due to limitation of moisture availability.

13.3 Summary

Climate change impacts the rate of evaporation from open water and evapotranspiration from vegetation. It has been demonstrated that increasing solar radiation and increasing temperature increases evaporation and evapotranspiration. During drought periods, a deficit in rainfall results with more clear sky days and low humidity. This condition creates a highly favorable environment for increasing evaporation and evapotranspiration resulting in accelerated water loss from lakes, reservoirs, and the soil. Studies on climate change impact on hydrology of every region need to include changes in evaporation and evapotranspiration. Increase in CO_2 and available energy would increase plant productivity, if moisture is available. Plant reaction to climate change in the rate and amount of water use is a subject for study.

Acknowledgments The authors acknowledge Shimeles Setegn for providing climate model outputs for the Great River basin of Jamaica that was used to generate Figs. 3.2 and 3.3.

References

Abtew W (1996) Evapotranspiration measurements and modeling for three wetland systems in south Florida. Water Resour Bull 32(3):465–473

Arnold JG, Srinivasan R, Muttiah RR, Williams JR (1998) Large area hydrologic modeling and assessment – Part I: model development. J Am Water Resour Assoc 34(1):73–89

Azam F, Farooq S (2005) Agriculture and global warming: evapotranspiration as an important factor compared to CO2. Pak J Biol Sci 8(11):1630–1638

Bates B, Kundzewicz ZW, Wu S, Palutokof J (2008) Climate change and water. IPCC technical paper VI. UNEP, Geneva

Calanca P, Roesch A, Jasper K, Wild M (2006) Global warming and the summertime evapotranspiration of the Alpine region. Clim Change 79(1–2):65–78

California Department of Water Resources (2006) Progress in incorporating climate change into management of California water resources. Technical Memorandum Report. California Department of Water Resources, Sacramento

Chattopadhyay N, Hulme M (1997) Evaporation and potential evapotranspiration in India under conditions of recent and future climate change. Agric For Meteorol 87(1):55–73

Cong Z, Yang D, Lei Z (2008) Did evaporation paradox disappear after the 1980s? A case study for China. Geophysical Research Abstracts. EGU, Munich, Germany

Dai A, Trenberth KE, Qian T (2004) A global dataset of Palmer drought index for 1870–2002: relationship with soil moisture and effects of surface warming. J Hydrometeorol 5(6): 1117–1130

Goyal RK (2004) Sensitivity of evapotranspiration to global warming: a case study of arid zone of Rajasthan (India). Agric Water Manage 19(1):1–11

Hatfield J, Boote K, Fay P, Hahn L, Izaurralde C, Kimball BA, Mader T, Morgan J, Ort D, Polley W, Thomson A, Wolfe D (2008) Agriculture. In: Walsh M (ed) The effects of climate change on agriculture, land resources, water resources, and biodiversity. A Report by the U.S. Climate Change Science Program and the Subcommittee on Global Change Research, Washington, DC

International Panel on Climate Change (IPCC) (2007) Summary for policymakers. In: Climate change 2007: The physical science basis. Contribution of working group I to the fourth assessment report of the intergovernmental panel on climate change. In: Solomon S, Qin D, Manning M, Chen Z, Marquis M, Averyt KB, Tignor M, Miller HL (eds) Cambridge University Press, Cambridge, United Kingdom

Jung M et al (2010) Recent decline in the global evapotranspiration trend due to limited moisture supply. Nature 467:951–954

Melesse A, Setegn S, McClain M, Vicioso F, Veloz F, Fu J, Webber D, Ortiz J, Nunez F (2011) Caribbean coastal scenarios project, fifth year final report submitted to IAI under the agreement CRN II-061. Department of Environmental Studies, Florida International University, Miami, FL, May 2011, p 104

Obeysekera J, Park J, Ortiz MI, Trimble P, Barnes J, VanArman J, Said W, Gadzinski E (2011) Past and projected trends in climate and sea level for south Florida. Technical report. South Florida Water Management District, West Palm Beach, FL

Parry ML, Canziani OF, Palutikof JP, van der Linden PJ, Hanson CE (ed) (2007) Climate change 2007: impacts, adaptation and vulnerability. Contribution of working group II to the fourth assessment report of the intergovernmental panel on climate change (IPCC AR4 WG2 (2007). Cambridge University Press, Cambridge. ISBN 978-0-521-88010-7 (pb: 978-0-521-70597-4)

Twilley RR, Barron EJ, Gholz HL, Harwell MA, Miller RL, Reed DJ, Rose JB, Siemann EH, Wetzel RG, Zimmerman RJ (2001) Confronting climate change in the Gulf Coast Region. A report of the Union of Concerned Scientists and the Ecological Society of America. http://www.ucsusa.org/assets/documents/global_warming/gulfcoast.pdf. Accessed 6 Aug 2012

Vehviläinen B, Huttunen M (1997) Climate change and water resources in Finland. Boreal Environ Res 2:3–16

Zhang M, Geng S, Ransom M, Ustin S (1996) The effects of global warming on evapotranspiration and alfalfa production in California. Department of Land, Air and Water Resources, University of California, Davis

Index

W. Abtew and A. Melesse, *Evaporation and Evapotranspiration: Measurements and Estimations*, DOI 10.1007/978-94-007-4737-1,